A SHEARWATER BOOK

Under
Ground

A Project of SCOPE,
the Scientific Committee
on Problems of
the Environment

Yvonne Baskin

How Creatures

of Mud and Dirt

Shape Our World

 Under
Ground

 **ISLAND**PRESS

SHEARWATER BOOKS

WASHINGTON • COVELO • LONDON

A Shearwater Book
Published by Island Press

Copyright © 2005 The Scientific Committee on Problems of the Environment (SCOPE)

Shearwater Books is a trademark of The Center for Resource Economics.

Library of Congress Cataloging-in-Publication data.

Baskin, Yvonne.
  Under ground : how creatures of mud and dirt shape our world / Yvonne Baskin.
    p. cm.
  Includes bibliographical references and index.
  ISBN 1-59726-003-7 (cloth : alk. paper)
  1. Soil animals. 2. Burrowing animals. I. Title.
  QL110.B35 2005
  591.75′7—dc22

                                                                    2004030330

British Cataloguing-in-Publication data available.

Printed on recycled, acid-free paper ✷

Design by McKnight Design, LLC

Manufactured in the United States of America

10 9 8 7 6 5 4 3 2 1

Contents

I

INTRODUCTION

Opening the Black Box

Two golf cart–sized rovers named Opportunity and Spirit bounced to a landing on opposite sides of Mars in early 2004. From 200 million miles away, NASA scientists sent these robotic vehicles rolling about the rubble-strewn surface, poking their sophisticated instrument-tipped arms at rock outcrops, dunes, and dusty plains. Their mission: to search for geologic evidence that Mars was once a warmer, wetter, and perhaps even habitable planet.

The prospect of life on Mars has captivated dreamers and visionaries for ages. Barely a century ago, astronomers and fantasy writers could peer into the night sky and imagine the red planet's mottled surface laced with canals or seething with warlike aliens set to invade Earth. In the 1960s, the first images beamed back to us by Mariner spacecraft quashed any lingering visions of canals or ruined cities. If we were ever to find signs of Martian life, it was clear we would have to search beneath the surface of an arid, bitterly cold planet with air too thin to breathe. A Viking lander did just that in 1976: it scooped up material from the planet's surface, analyzed it chemically, and found no clear evidence of life. That disappointment, however, did not quench our curiosity. Perhaps there was once a golden age on Mars, a warmer time when the planet nodded toward the Sun, polar ice melted, rivers flowed, seas surged, and life took hold.

It was almost three decades later when ecstatic space agency scientists announced that Opportunity had found evidence that Mars once hosted water—not just soggy plains but a shallow equatorial sea or swamp of briny, acidic water that had left ripple patterns and salt deposits in surface rocks.[1] For months I followed the news and watched online as the rovers beamed back startlingly clear images of the Martian surface.

Captivated as I was, one seemingly trivial point kept jarring me. News reports and often the scientists themselves persisted in calling the loose stuff that the rovers were probing "soil."

Soil? Among space buffs, that use of the word had become common enough that *Merriam-Webster's Collegiate Dictionary* defines soil in one sense as "the superficial unconsolidated and usually weathered part of the mantle of a planet."[2] But by the time Spirit and Opportunity were sent roving across Mars, I had spent more than a year learning about the mysteries of the earthly stuff we call soil, and using that word to describe the Martian surface sounded, well, oddly alien. Consider this description from Daniel Richter of Duke University who, like many ecologists, considers soil to be not simply the loose surface material of a planet but "the central processing unit of the earth's environment":

> Soil is the biologically excited layer of the earth's crust. It is an organized mixture of organic and mineral matter. Soil is created by and responsive to organisms, climate, geologic processes, and the chemistry of the aboveground atmosphere. Soil is the rooting zone for terrestrial plants and the filtration medium that influences the quality and quantity of Earth's waters. Soil supports the nearly unexplored communities of microorganisms that decompose organic matter and recirculate many of the biosphere's chemical elements.[3]

In this light, Mars enthusiasts are jumping the gun when they call the dust of that planet "soil." Theirs is an understandably hopeful act—a hope that the Martian surface contains, if not current life, at least a legacy of life. So far, however, such hopes linger unfulfilled. True soil, as ecologists see it, remains at least as rare in the universe as life itself. Indeed, life—abundant and long-flourishing life—must precede soil. It is life that substantially organizes and transforms the weathered parent material of the planet into soil. The only soil discovered so far is often called "earth" after the only planet on which it's found.

Ironically, the money and vision expended on probing the secrets of Mars—$820 million for the latest two rovers alone—vastly exceed what has been spent exploring the earth beneath our feet. Yet it is the soils of our gardens, fields, pastures, and forests, as well as the sediments beneath streams, lakes, marshes, and seas, that harbor the most diverse and abundant web of life known in the universe. What's more, it is life underground that makes possible the green and fruitful surface world that allows us to create flourishing civilizations with the means and the curiosity to probe the universe.

Although money for exploring soil life remains relatively sparse, the pace of exploration and sense of excitement are growing among scientists who look down instead of up. Like space scientists, soil ecologists, too, are harnessing new technologies to reveal cryptic realms as little understood as the rusty skin of Mars—and far more vital to our existence. Unlike space exploration, however, the drive to understand life underground is fueled by a sense of urgency. Human activities are increasingly degrading and impoverishing soils and soil life, and this loss, in turn, threatens to diminish the earth's capacity to sustain us.

Soils have been called "the poor man's rainforest" because a spade of rich garden soil may harbor more species than the entire Amazon nurtures aboveground.[4] Two-thirds of the earth's biological diversity— biodiversity for short—lives in its terrestrial soils and underwater sediments, a micromenagerie that includes uncataloged millions of microbes, mainly bacteria and fungi; single-celled protozoa; and tiny animals such as nematodes, copepods, springtails, mites, beetles, snails, shrimp, termites, pillbugs, and earthworms. Some are accessible to anyone curious enough to poke through rotting leaves, backyard dirt, or the muddy bottom of a tidal marsh, but most are too small to see without a microscope or magnifying glass. So little effort has been devoted to life underground—and so few scientists specialize in identifying these organisms—that at best only 5 percent of the species in most key groups of soil animals have so far been identified,[5] and in marine sediments, less than 0.1 percent of species may be known.[6]

Taken together, however, these inconspicuous creatures dominate life on earth, not just in diversity but also in sheer numbers and even body mass. Harvard University ecologist Edward O. Wilson points out that 93 percent of the "dry weight of animal tissue" in a patch of Amazonian rain forest in Brazil belongs to invertebrates living everywhere from soil to treetops, from mites and springtails to ants and termites.[7] And that doesn't count the microbes. Despite their submicron stature, the bacteria in an acre of soil can outweigh a cow or two grazing above them.[8] Indeed, bacteria may contain more than half of the "living protoplasm" on earth, most of it to be found either in terrestrial soils or in the mud of the oceans that cover three-fourths of the planet.[9]

Underworld creatures are not only numerous and weighty in aggregate, but ancient and exceedingly durable. Toughest among them are the "extremophiles," bacteria and ancient microbes known as archaea that can live a mile or more deep in the earth, or in boiling hot springs or polar ice, enduring extremes of heat, cold, pressure, and pH that were considered unfailingly lethal to any form of life only a few decades ago.[10] Some tiny soil animals can time-travel for decades or more in dormant states, impervious to extreme heat, cold, desiccation, and otherwise lethal radiation.[11] Although most soil organisms are small and short-lived, some of the oldest and largest creatures ever identified are sprawling underground masses of the root-rot fungus *Armillaria* that far outclass blue whales in size. A 220,000-pound specimen that stretches across 37 acres of Michigan woodland was

Two-thirds of the earth's biological diversity lives in its soils and underwater sediments, and thriving underground communities keep the planet's surface green and habitable.

reported in 1992, setting off a race of sorts to find the biggest "humongous fungus."[12] By 2003, a 2,200-acre *Armillaria* in Oregon had captured the record.[13] Finally, although we think of plants as denizens of our aboveground world, many plants spend more than half the energy they capture from the sun to grow roots that nurture and interact with life underground.[14] A prairie, for example, grows more grass biomass below the surface than above.

If scientists still know very little about who lives underground, they know even less about what each species in particular does for a living. Yet the creatures of mud and dirt are so important to our life that Wilson calls them "the little things that run the world."[15] Together they form the foundation for the earth's food webs, break down organic matter, store and recycle nutrients vital to plant growth, generate soil, renew soil fertility, filter and purify water, degrade and detoxify pollutants, control plant pests and pathogens, yield up our most important antibiotics, and help determine the fate of carbon and greenhouse gases and thus, the state of the earth's atmosphere and climate. All of these ecological services arise from the spontaneous activities of billions of creatures going about the business of nourishing and reproducing themselves in a series of elaborate food webs below the surface.

Since the dawn of agriculture, humans have recognized the value of the soil itself, often invoking its fertility in ritual and sacrifice. Yet most societies have given little thought to, or have been simply unaware of, the multitude of creatures that live and work in the soil. The scientific study of soil developed in the 19th century, driven largely by the desire for greater crop production. Even soil scientists, however, have traditionally treated the soil as a "black box"—a system whose internal workings remain hidden or mysterious—measuring physical and chemical attributes such as pH and organic matter content, monitoring inputs of nitrogen and outputs of carbon dioxide, but making little effort to identify the dynamic workforce within. Yet we now know that these soil attributes and outputs reflect the legacy of billions of organisms eating, breathing, growing, interacting with one another, and, in the process, altering their environment—and ours.

Today, a growing cadre of scientists drawn from numerous disciplines and armed with new techniques is working to crack open the black box of soil life and soil processes and fill in that sketchy outline with deeper understanding. Soil ecologists in the 1950s pioneered research on soil biodiversity, food webs, and soil-plant interactions, but since the 1980s that effort has burgeoned dramatically in parallel with the development of ecosystem science.[16] Researchers today view soils and sediments as complex ecosystems, and they recognize that the processes that take place underground vitally affect not only our food and timber supplies but also the quality and sustainability of our environment. Soils and aquatic sediments now draw the attention of multidisciplinary teams of, for example, ecologists, biogeochemists, microbiologists, zoologists, entomologists, agronomists, foresters, marine and freshwater biologists, geologists, and atmospheric scientists. These researchers want to know who is down there, what each contributes to the functioning of the soil, how they are organized into communities and food webs, why some communities are richer in species than others, and how our activities threaten soil life and processes.

Unlike Mars exploration, the increasing effort to understand life underground is not driven by curiosity or futuristic speculation alone. The diversity of life in soils and sediments is under increasing threat, just like plant and animal life aboveground, and as a result so is the integrity of the ecological processes that are influenced by underground life.

By some estimates, more than 40 percent of the earth's plant-covered lands, from dry rangelands to tropical rain forests, have been degraded over the past half-century by direct human uses such as grazing, timber cutting, and farming. Degraded land, by definition, has a diminished capacity to grow crops and forests and supply other goods and life support services to humanity.[17] In that same half-century, erosion has lowered potential harvests on as much as 30 percent of the world's farmlands. Erosion not only sweeps away mineral soil but also reduces the abundance and diversity of soil creatures, which are concentrated in the top few inches of the soil. "A hectare [2.5 acres] of good quality soil contains an average of 1,000 kg [kilograms— 2,200 pounds] of earthworms, 1,000 kg of arthropods, 150 kg [330 pounds] of protozoa, 150 kg of algae, 1,700 kg [3,740 pounds] of bacteria, and 2,700 kg [5,940 pounds] of fungi," according to Cornell University ecologist David Pimentel.[18] As this life is lost, the soil's ability to hold water and nurture crops declines. Further, as soil and nutrients wash off the land and into rivers, lakes, and coastal waters, they damage water quality and smother and degrade sediment communities often already disrupted by pollution, dredging, and trawl fishing. Human-driven changes in climate, acid rain, excessive nitrogen deposition, the spread of nonnative species, and the continuing conversion of land to crops, cities, and other human uses all contribute to the loss of soil biodiversity and functioning.[19]

Accelerating degradation of the earth's soils and sediments has not gone unnoticed by national and international organizations concerned with agricultural productivity, fisheries, food security, and poverty relief as well as biodiversity.[20] Increasingly they recognize that defining, preserving, and restoring the health of soils and sediments are fundamental to addressing such problems as climate change, desertification, declining water quality, and the sustainability of agriculture, forestry, and fisheries worldwide. In turn, the health and quality of soils and sediments rely fundamentally on the work of the living communities within them.

One of the international efforts that grew out of this concern is the Soil and Sediment Biodiversity and Ecosystem Functioning project led by soil ecologist Diana Wall of Colorado State University and sponsored by a nongovernmental scientific organization known as the Scientific Committee on Problems of the Environment (SCOPE). Since 1996, a wide array of specialists from around the world has volunteered time to the project to pull together what is known about the biodiversity of the earth's soils and freshwater and marine sediments, its role in sustaining vital ecological processes, and threats to soil organisms and the services they provide. This book is an outgrowth of that project, and access to participating scientists has allowed me to explore how human activities threaten the integrity of soil and sediment communities, and in turn, the critical services they provide to human society.

The idea for this book grew out of a chance encounter in February 2001 when I happened upon Wall and John W. B. Stewart, a retired soil scientist and SCOPE editor in chief, outside a hotel conference room in San Francisco during a scientific meeting. I had already written one book based on findings from a SCOPE project and at the time was writing a second.[21] Wall began telling me about the soil and sediment project and asked if I would like to get involved. How could I not be interested? Her enthusiasm for her science is infectious, and I'm an obsessive gardener, at least during the brief months when the soils in southwest Montana thaw. Furthermore, I had become fascinated by the link between biodiversity and ecological processes while working on my first SCOPE-sponsored book in the early 1990s. So little was known at that time about the ecological roles of specific soil creatures that SCOPE decided to launch a new effort—Wall's project—focused specifically on soil and sediments. The first question that occurred to me was would I be able to learn enough about soil life from the results of this second effort to fill a whole book? Wall assured me I would, and she followed up in the months ahead with stacks of journal articles and reports the project teams had produced. That material introduced me to a topic much larger and more significant than I had imagined.

Almost 2 years later, in November 2002, I joined more than two dozen project scientists who had gathered at a lodge in Estes Park, Colorado, to synthesize what they had learned about soil and sediment biodiversity, its vulnerability to human activities, and strategies for its future conservation and management.[22] That was my first opportunity to mingle with people who "see" below the surface and are aware of and concerned about the underground world. I began to probe for details, to look for situations and stories that would illustrate the work of soil communities and their great relevance to our own well-being.

From Estes Park, my explorations of life underground took me to the polar desert of Antarctica, the coastal rain forests of Canada, the rangelands of Yellowstone National Park, the vanishing wetlands of the Mississippi River basin, Dutch pastures, and English sounds. This was not a journey of lament through ruined landscapes but an opportunity to walk and talk with scientists and land managers who are pioneering ways to integrate new knowledge about soil life into efforts to restore, sustain, or monitor the health of our lands and waters. In this book you will hear from a marine ecologist who monitors the work of burrowing shrimp in Plymouth Sound in hope of gaining more protection for important mud-bottom creatures everywhere in the debate over acceptable fishing practices; learn about researchers in England and the Netherlands who are trying to reverse degradation caused by intensive agriculture on former croplands; follow Canadian forest ecologists as they explore the fate of root fungi vital to forest regeneration in stands logged using controversial "new forestry" techniques; and join ecologists tracking the destructive advance of exotic earthworms through a Minnesota sugar maple forest.

The result is not a comprehensive tome on soil ecology but a series of windows to an unseen world that is fascinating in its own right, vital to our well-being, and yet increasingly threatened by our activities. Where possible, I introduce you to the lives and significance of specific creatures or groups of creatures in hopes that you will begin, as I have, to marvel at and perhaps respect the world underground. I have chosen to portray the workings of soil life not in the familiar settings of our lawns and gardens but in contexts that I found unexpected and sometimes startling. My message is that creatures of the mud and dirt lead larger lives and shape the world we experience more powerfully than most of us imagine. Their first service, in fact, was to transform Earth into a planet suitable for life.

Some 4.5 billion years ago, swirls of hot interstellar gases and dust began coalescing to form Earth and our solar system.[23] For hundreds of millions of years thereafter, massive chunks of rock or ice continued to batter our young planet, periodically melting its crust or boiling away the warm oceans that formed in million-year torrents as the planet cooled. By 3.9 billion years ago, those collisions had grown rare and continents began to rise. Earth was still hot, its atmosphere devoid of free oxygen and lacking a protective ozone layer that could buffer the molecule-shattering ultraviolet radiation from the young sun. Somewhere on the planet, however, life was in the making—perhaps in warm shallow coastal waters, in the open ocean, in hydrothermal vents bubbling from the seafloor, or even deep underground. Wherever it arose, though, this early life itself helped transform Earth into the uniquely habitable planet we enjoy today.

By 3 billion years ago, communities dominated by mats of cyanobacteria thrived in the shallow waters of the planet. Cyanobacteria —once called blue-green algae—are ubiquitous in the earth's soils and waters today, visible in forms ranging from pond scum to living crusts on the desert floor. They pull the nitrogen they need directly from the air and also make their own food through photosynthesis just as green plants do. Using sunlight for energy, these single-celled creatures breathe in carbon dioxide, strip the carbon from it, and use the carbon to assemble sugars and other organic compounds needed to build and fuel life. In the process, the microbes discard the oxygen molecules from the carbon dioxide, creating what paleontologist Richard Fortey calls "the most precious waste in the firmament."[24] Over a billion or so years, the exhalations of microbes created the earth's oxygen-rich atmosphere and protective ozone layer that allowed more complex life to evolve.

Ancient microbes probably transformed the land surface as well as the air. At some point, cyanobacteria and other microbes emerged from the shallow waters onto the inhospitable shores, forming themselves into rich slimes, mats, and crusts that protected them from drying. The organic acids these one-celled life forms secreted helped to speed the weathering of parent rock to sand, silt, and clay and added organic matter to the nascent soil. The sticky slimes would have stabilized this loose material against erosion and allowed the first soils to accumulate.[25]

With the so-called Cambrian Explosion 530 million years ago, animal life came into its own, arising and proliferating in the waters and muck of the seafloor. Some 400 million years ago, the descendents of that explosion began to emerge onto land. In the vanguard were the ancestors of many of today's underground dwellers—tiny flatworms, springtails, mites, pseudoscorpions, spiderlike creatures, and the scurrying predecessors of modern insects (many of which live part or all of their lives underground). By 350 million years ago, the first green plants arose, and together with microbes and animals helped to drive creation of the vital soil systems we rely on today. As the roots of plants and their microbial followers pushed ever deeper into the soil, the carbon dioxide they exhaled reacted with rainwater, creating acids that helped to weather the rock of the earth's crust into sand, silt, and clay minerals. Those minerals, combined with air, water, and organic matter from decaying plant and animal material, along with living organisms, are the key constituents of soil.[26] It takes hundreds to thousands of years to create soil from rock, depending on its hardness; sandstone or shale clearly yields faster than granite. The process of soil formation is so slow relative to the human lifespan that it seems unrealistic to consider soil a renewable resource. By one estimate, it takes 200–1,000 years to regenerate an inch of lost topsoil.[27] That is one reason both ecologists and agronomists become alarmed at farming or construction practices or other human activities that promote excessive erosion of topsoil.

Scientists classify the earth's soils, like its life forms, into an intricate and constantly shifting taxonomy. There are 11 major orders of soil, from the dark, fertile Mollisols of temperate grasslands to the highly weathered yellow Oxisols of the humid tropics. Within these orders are numerous subcategories encompassing tens of thousands of distinct soil series worldwide, more than 13,000 in the United States alone. Each soil series is equivalent to a biological species, and the "profile" of its horizontal layers or "horizons" represents a unique interaction of climate and life with parent rocks and topography in a specific place through time. The result is a soil with unique texture, structure, organic matter content, and living communities.[28] In turn, the character of the soil helps determine whether we encounter fir forests, grassy savannas, or sagebrush above, and whether the land can be converted to grow wheat or tomatoes or oranges.

Until recent decades, soil science focused primarily on agriculture, and only the organic-rich upper horizons to the depth of crop roots were considered soil. Now the definition of soil is being pushed ever deeper into the earth by scientists concerned with everything from the influence of deeply rooted plants and deep-dwelling microbes to groundwater supplies and the fate of pollutants. Some disciplines define the lower limit of the soil at about 6 feet, whereas others see the zone of biological influence extending 30 feet or even hundreds of feet into the earth's crust.[29] Increasingly, scientists recognize that life deep underground can influence everything from the quality of our water supplies to the character of life aboveground.

If more effort has in the past been spent classifying the soils of the earth than examining the work of the living communities within, that is changing rapidly, and the modern efforts to shed light on the black box of the soil are the focus of this book. Paradoxically, the below-ground life that we have long ignored or taken for granted is not only more important for our survival, but arguably as bizarre and alien as anything we are likely to find in the dust, ice, or seas of another planet. It seems fitting then to begin the story of life underground with a visit to scientists who are probing the soils of the most Mars-like place on our planet, a continent once lush and temperate until geologic forces drove it into its present position at the end of the Earth.

II 🪲

Where
Nematodes
Are Lions

On a brilliant mid-summer day in December, our helicopter lifts off from McMurdo Station, the largest out-
post of the U.S. Antarctic Program, situated some 2,400
miles south of New Zealand. A quick 50-mile flight
across frozen McMurdo Sound brings us to a dark
rocky beach at the mouth of Taylor Valley, the southernmost of the Mc-
Murdo Dry Valleys. These valleys are a unique creation of the
Transantarctic Mountains, which form an 1,800-mile-long spine separating East from West Antarctica and block the advance of the massive
East Antarctic ice sheet toward the sea. In a handful of valleys bordering McMurdo Sound, fierce scouring winds conspire with the bul-
wark of the Transantarctic ridges to create the largest ice-free expanse
on a continent largely frozen for 30 million years. The polar deserts of
the dry valleys are often touted as the most Mars-like terrain on Earth.

Turning up Taylor Valley, we fly over a landscape of glacial rubble
patterned into tortoiseshell polygons by the heaving and sighing of
frozen ground. Along the valley wall to our right, glacial tongues lap
out between peaks of the Asgard Range, descending to the valley floor
and coming to a halt as stark, blue-white ice walls that rise as high
as 65 feet above the bleak terrain. We pass the Commonwealth Glacier

and advance toward the Canada. Silver ribbons of meltwater stream from these and smaller glaciers, meandering across the valley floor until they disappear into a liquid moat that rings the permanent ice cover of Lake Fryxell below us.

The Taylor Valley glaciers and lakes—first Fryxell, then Lake Hoare, and at the head of this 22-mile-long valley, Lake Bonney—look almost insignificant from the air. But set off to hike among them and you soon realize that the clear air and stark landscape fool the eye. There are no trees, no familiar living shapes to help judge size and distance, no sound but wind. That very starkness, however, is the reason the research team I'm flying with returns here each December at the peak of the austral summer.[1]

Despite their barren appearance, the dry valleys serve as an oasis for land-based life on a continent 98 percent concealed by ice. Below us, life persists largely unseen in the soils, rock, ice, and streambeds and also in permanently liquid stews of briny water beneath the lake ice. This is a sparse world, largely microbial, but with a smattering of microscopic invertebrate animals to round out a simplified food chain. In the early 1990s, scientists from many disciplines began converging on this stripped-down ecosystem each summer in a coordinated effort to decipher ecological patterns and processes too complex to unravel in livelier, greener places.[2]

"This is the only place where we can see the effect of a change or disturbance on an individual species in the soil," soil ecologist Diana Wall had told me a week earlier as we waited in Christchurch, New Zealand, for the military cargo plane that would ferry us across the Southern Ocean to McMurdo. "We want to know how human-caused changes in climate could influence members of the soil food web, and what effect the loss of individual soil species might have on ecological processes such as nutrient cycling," Wall said.

Now, as our helicopter banks to land near the shore of Lake Fryxell, Wall can barely contain her excitement. She points down toward rows of translucent plastic cones glinting like lampshades on a nearby slope. Director of the Natural Resource Ecology Laboratory at Colorado State University, Wall is coleader of a research team long

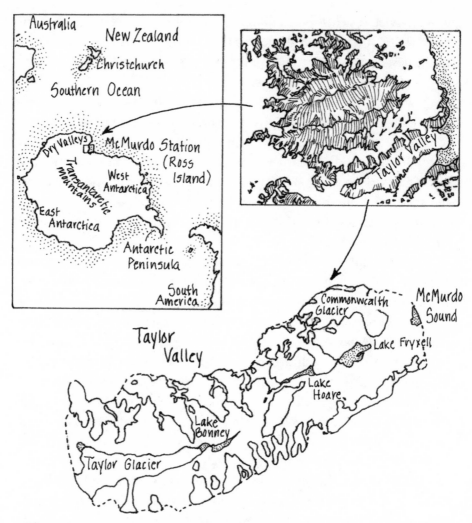

The dry valleys bordering McMurdo Sound provide a refuge for land-based life on the largely ice-bound Antarctic continent.

known here as the "Wormherders" because their efforts focus chiefly on the fortunes of nematodes, microbe-munching roundworms about 1/20th of an inch long that dominate the food chain of the dry valleys like lions on the savanna. The field of cones—actually, cone-shaped warming chambers—is one of the "worm farms" we've come to tend.

It is fortuitous for Wall that the animals she has studied for more than three decades dominate these valleys, but it is hardly surprising. Nematodes are the most diverse and abundant animals on the planet, outnumbering even ants. Four of every five animals are nematodes.[3] These mostly microscopic and transparent creatures live in our gardens and crop fields, in oceans and lakes, inside the bodies of bees and horses, whales and us. Most of the free-living nematodes in soils and sediments graze on bacteria, fungi, and algae, breaking down the organic matter tied up in these microbial hordes and speeding nutrient cycling by releasing key building blocks of life such as carbon and nitrogen that will nurture and fuel new generations of beings above and below the ground.

In complex soil food webs on other continents, nematodes graze amid protozoa, slime molds, springtails (wingless relatives of insects in the animal phylum or grouping known as arthropods), and other invertebrates that also consume microbes. Predators such as mites, pseudoscorpions, centipedes, and spiders feed on the grazers, and in turn serve as food for larger predators. This complexity masks the importance of any single species in the vast business of nutrient cycling. Thus, the very sparseness of the soil food web in Antarctica makes this an attractive place to explore one of the most urgent questions in ecological research: What do we lose in terms of ecological functioning as species disappear?

"Here we have a group of animals in an extreme environment who are involved in decomposition and nutrient cycling just like their peers in other soil ecosystems," Wall explained. "So it's not a stretch of the imagination to take this animal living in the soil in Antarctica, subject it to climate change or other disturbance, and predict that this is what might happen elsewhere."

For more than a decade, Wall and her collaborators have been altering the temperature, moisture, and food supplies inside the worm farm chambers to see how nematodes respond. Ironically, the climate of the dry valleys has subjected nematodes to an even more severe test during this period, and populations have plummeted.

The pilot touches our helicopter down on a flat square of sand outlined with rocks near a blue hut, the headquarters for field camp F6. Three of us pile out, crouching low under the still-turning rotors, and drag a bevy of gear-filled ice chests, plastic buckets, daypacks, and dozens of 5-gallon carboys full of water and sugar solutions safely beyond the propeller wash before the helicopter lifts off.

Antarctica is billed, without exaggeration, as the highest, driest, coldest, windiest place on earth. You don't have to wait long after arriving on "the ice," as everyone here calls the continent, to hear lurid tales of chilling deaths, and not just among the hoary explorers of a century past. Only a few days earlier I'd completed a mandatory overnight survival school with Wall's two postdoctoral researchers—all three of us new to the ice—and listened to cautionary tales of folly as we dug snow shelters, learned to use two-way radios, and practiced rescuing one another in a simulated whiteout. Although Taylor Valley is ice-free and averages less precipitation than the Sahara Desert—a scant 4 inches of snow a year, the equivalent of a fraction of an inch of rain—the mean annual temperature ranges from 3° to −6° F, and the winds that sweep down off the ice sheet or up from the sea can quickly drop the wind chill as low as −100° F.

On this morning, however, the sun is piercing and the temperature hovers in the high 20s. Without the brisk wind, the place would seem almost balmy for three people freshly arrived from winter in the Rocky Mountains: Wall from Colorado; Byron Adams, an evolutionary biologist from Brigham Young University in Utah; and me from Montana. We hurry to adjust clothing layers and lace up our hiking boots. Wall is moving quickly, and her sense of excitement and haste on the first field visit of the season is infectious. Time is critical, both for us and for the soil life we've come to monitor. The soil community here endures in suspended half-life through the long, dark polar winter, waiting for golden days like this one each December and January. In this brief polar summer, the sun shines round the clock, the air temperature rises near freezing, the soil surface absorbs enough solar radiation to thaw, glaciers melt a bit, and liquid water brings streams to life. For the scientists who come here, helicopter hours are rationed and field time crowded with repetitive and often exhausting tasks. I quickly learn that little of the soil team's fieldwork is high tech. We will spend the next 5 hours emptying dust traps; scooping up, bagging, and labeling soil; and pouring water and sugar solutions on some long-pampered little communities sheltered inside the chambers of the worm farm.

I hustle to keep up as Wall and Adams grab packs and buckets and move quickly up the slopes beyond the hut, following a well-worn path past a scattering of blue and yellow sleeping tents and, when the path ends, walking through what feels like the deep sand and jumbled rock of a dry streambed.

"Geez, it's awesome; this place is great." Adams, irrepressibly cheerful, is admiring the glaciers and peaks above us as he hurries toward a line of dust traps. This is his second season on the ice. I gawk up at the stunning scenery, too, while trying to keep my footing on the uneven ground.

"I try not to make new footprints," Wall says matter-of-factly, walking carefully along the troughs of the polygon-patterned ground. "There's so much traffic out here now." I take that as a subtle caution and focus on my feet, seeing only rubble where she sees an ecosystem. Indeed, the Wormherders have learned that the centers of the polygons offer the best habitat and host the most abundant nematode populations.[4] I try to keep my feet in the narrow troughs that define the polygon boundaries.

At first, Wall seems an unlikely person to be found kneeling for hours in the dirt, hands cracked and bleeding from digging gloveless in near-frozen ground. Students and colleagues fondly describe her as a "type specimen" of an overachiever. Besides directing a major research center and leading multiple international collaborations, Wall has presided over a growing list of professional societies and international panels, committees, and programs that keep her jetting around the globe much of the year. Yet Wall had been captivated by the romance of Antarctic exploration for 20 years before her research interests presented her with a reason to make the journey herself. From her first season on the ice in 1989, she has remained fiercely devoted to the place and its science, returning annually to the dry valleys with her team as well as longtime collaborator Ross Virginia of Dartmouth College or members of his research group. The team quickly learned that life persists perilously close to the edge here, and that, as Wall puts it, "every human footprint is an ecological footprint."

During the 1990s, the Antarctic Treaty nations acknowledged the fragility of the dry valleys by designating them as a special management area and adopting regulations to protect them from pollution, human waste, vehicles, and even footprints where possible. These may be the only soil and sediment communities in the world with such protections, and it explains why we are all carrying "pee bottles" in our packs and why the helicopters that pass overhead are often "slinging" 50-gallon drums of human waste from the field camps back to McMurdo. (The Antarctic Treaty, first signed in 1959, declares that no country owns or rules Antarctica and that the continent is to be dedicated to peaceful purposes such as scientific research. Some 44 nations are now parties to the treaty.)

As the day proceeds, I make myself useful by holding open twist-top plastic bags while Wall carefully pours wind-blown soils from red nylon trays that have been sitting out in open-topped chambers all winter. The trays serve as dust traps that are helping the team test a theory that winds sweep dormant nematodes and other tiny invertebrates and microbes around the valleys and even disperse them far onto the continent.

Wall picks up a palm-sized rock that had been placed as a weight in the center of one tray. The wind has sandblasted its black top into soft curves. The bottom, buried for ages in the soil, is stained a lighter color. Under rocks like this that pave the polygon surfaces, in ancient soils as coarse-textured as beach sand and often salty and alkaline, live single-celled green algae, cyanobacteria, microbes, and other invertebrate animals as well as nematodes.

"I just hate that we move these," Wall says as she puts the rock aside. "There was a community under this rock that took thousands of years to form."

She lifts the tray over the plastic bag I'm holding out. "Okay, let's see where you're from," she says, talking to the creatures she envisions in the accumulated dust as she pours. We won't know who is actually there until we return to the McMurdo lab, flush them from the soil with sugar and water, and examine them under a compound microscope.

"We have so many questions we want to ask now," Wall explains. "But our first years here were very much a discovery process. First we just wanted to find out if these beasts were here."

Her comments remind me that until she and Ross Virginia began their fieldwork, few people believed there were any living creatures out here at all. In and around the lakes and streams, yes, but not out here in the arid soils that cover 95 percent of the dry valleys. This soil was considered as sterile as the dust of Mars or the moon.

British explorer Robert Falcon Scott and two companions became the first people to set foot in the dry valleys when they descended into Taylor Valley from the eastern ice sheet in December 1903. After the hardships of the polar plateau, the party delighted in the novelty of lunching on a sandy beach beside a gurgling stream. The only sign of life they noticed was the skeleton of a Weddell seal that had inexplicably hauled itself 20 miles up from the sea. In his journal, Scott called this place the "valley of the dead."[5]

Scott was wrong, but he was hardly the last of us to overlook life right under our feet. Another 55 years would pass before anyone took a closer look. Interest in the biology of the valleys began with explorations conducted during the International Geophysical Year in 1957–1958, when researchers first documented a surprising array of life forms. At the edges of lakes and in ephemeral ponds and streams, researchers found mosses, lichens, and mats of green algae and red, orange, and black cyanobacteria. Living among the mats were bacteria, yeasts, molds, and an array of microscopic invertebrates that feed on microbes, algae, and detritus: nematodes, protozoa, rotifers (tiny aquatic invertebrates known as wheel animals because the beating of their hair-like cilia as they move and feed resembles a rotating wheel), tardigrades (chubby creatures variously nicknamed "moss pigs" or "water bears" because of the claws on their four pairs of stumpy legs), and occasionally, mites and springtails.[6]

Out beyond the watery habitats, however, investigators were drawing a blank in their efforts to detect even microbial life in the soils using the limited techniques of the day—primarily attempting to grow microbes on growth media and broths in Petri dishes, a method that reveals only 0.1–1 percent of the microbes in most soils.[7] Since no microbes could be found, biologists saw little reason to look for nematodes and other organisms that feed on microbes.

Much of the early biological research in the dry valleys involved scientists interested in the practical problems of searching for life on Mars. Because the arid soils appeared sterile and microbes from wetter habitats nearby had apparently failed to adapt and actively colonize the arid areas, some scientists suggested that "Martian life could not be built on a terrestrial model."[8] Earthly life seemed to have reached its limits here in conditions much less harsh than those on Mars. But other scientists weren't convinced. One member of the Viking mission biology team, Wolf Vishniac, fell to his death from a steep slope in the Asgard Range in 1973 while trying to disprove the sterility theory and develop a better "life detector" to send to Mars.[9] Vishniac and other skeptics were soon proven right: Life has learned to cope with conditions here.

The first direct sightings of life in the polar desert away from ponds and streams came in the mid-1970s, when E. Imre Friedmann and Roseli Ocampo reported finding cyanobacteria and later lichens—a partnership of green algae and fungi—growing within rock fissures and even in the pores of sandstone rocks in the mountains of the dry valleys region. Earlier, the two researchers had found similar "cryptoendolithic"—literally, "hidden in rock"—communities secreted within rocks a world away in hot deserts. In both places, it turned out, these microbes had adapted to aridity in the same way: When water becomes scarce, the organisms simply dry up, shut down metabolic activity, and wait in a "cryptobiotic" state until water again becomes available.[10] (Cryptobiotic translates literally as "hidden life," but it is used to describe various states of dormancy in which metabolic activity temporarily ceases and life is essentially suspended.) Similarly, the cyanobacteria and algae that form living crusts across the surface of many desert soils pass the dry periods in a dormant state, just like their cousins within the rocks.

Wall and Virginia, too, had done much of their research in hot deserts before they turned to the Antarctic. Working in the Chihuahuan desert of southern New Mexico, they had already learned that the diversity and abundance of nematodes are not tied to soil moisture levels. The finding seems counterintuitive because nematodes are essentially aquatic animals that live in water films on soil particles and in soil pores. The key to this paradox is that these tiny animals also have cryptobiotic strategies that allow them to shut down their life processes during dry spells.[11]

It seemed quite possible to Wall that nematodes could have colonized the arid soils of the dry valleys. But why, I wondered, with plenty of hot deserts to study, would a soil ecologist want to look for worms in Antarctica? One answer: to escape from the influence of plants.

In hot deserts, and indeed, in most other land-based ecosystems, green plants rather than water hold the key to where you will find the highest abundance and diversity of soil creatures. Shrubs such as mesquite create fertile islands, building up organic matter and nutrients around themselves thanks to their litter and roots. Even in the relatively barren stretches between mesquite shrubs, an underground network of mesquite roots exerts a powerful influence on the soil community. At some scale, patterns of underground life are also influenced directly by physical and chemical properties of the soil, but that influence—so stark on the frost-patterned ground of the dry valleys—is hard to detect amid the dominating presence of plants and the teeming activities of the soil communities around their roots.

Think of it this way: Life is not randomly scattered throughout the soil. Plant roots, leaf litter, animal burrows, termite mounds, earthworm castings, and other biological detritus as well as physical and chemical factors such as pH and salinity create a patchwork of good and poor neighborhoods underground.[12] The good neighborhoods are hotspots for diverse soil life and for the biological activities that drive decomposition, nutrient cycling, and other processes vital to plant growth. In many ecosystems, plants devote as much or more of the carbon they take in through photosynthesis to growing roots as to building new leaves and stems. Roots form a kind of upside-down forest, dominating the soil community with more than their physical presence. Growing roots push through the soil, drawing in water and soluble nutrients and at the same time sloughing dead cells and leaking significant amounts of sugars and amino acids into the "rhizosphere"—the neighborhood immediately adjacent to the roots. Microbes feast and flourish in the rhizosphere, growing tens or hundreds of times more numerous than microbial populations living in the bulk soil that often begins only 1/10th of an inch away. Protozoa, nematodes, and other consumers of microbes flourish, too, along with their predators and the rest of the soil food web. The rhizosphere is the place where symbiotic (mutually beneficial) interactions such as nitrogen fixation—a process by which microbes capture plant-fertilizing nitrogen from the air—take place, as does competition, predation, grazing, and other interactions between plants and the soil community.[13] Tree roots may plunge 25 feet or more, creating a three-dimensional ecosystem by moving carbon deep into the soil profile. Even the leafy canopies of plants alter the characteristics of the soil habitat by shading it and creating a layer of litter over the surface.[14]

"We got to thinking, what if you could just take the plant out of the system and have only the chemical and physical structure of the soil," Ross Virginia recalled one day as we sat in the third-floor library of the science lab building at McMurdo. "What would structure these nematode communities and how would they work?" This question is part of a larger mission to find out what individual soil species need, what they do in the soil, how they're vulnerable, and—more urgently, given the array of human threats to soil life—whether the loss of species can cause vital ecological processes to falter.

Most often, researchers approach such questions by using chemicals to knock out certain life forms—say, all plants or plant-feeding nematodes or all fungi—from a field plot. Or they resort to small-scale replicas called microcosms or mesocosms filled with sterilized soil to which they add manageable numbers of microbes, soil animal species, and perhaps plants. In the 1980s, Wall was using artificial systems such as these to look at how nematodes influence the movement of carbon through a system. At the suggestion of a colleague, she began to think about finding a real ecosystem that was not only naturally devoid of plants but harbored a limited number of soil animal species as well. Antarctica came to mind.

Wall contacted a colleague who was already working in Antarctica, and he mailed her three bottles of soil scooped from somewhere in Taylor Valley. She was able to extract a few nematodes from the samples, and on that basis, she and Virginia got their first grant to come to the ice.

Wall and Virginia already knew that Antarctica lacks higher plants, the green, rooted kind we're familiar with (except for a handful on the Antarctic Peninsula, which juts north above the Antarctic Circle). Soils here are nearly two-dimensional habitats, with most biological activity limited to the top 4 or 5 inches by the permanently frozen ground below. But before they turned from hot to cold deserts, Wall and Virginia needed to know just how much biological activity was actually taking place in Antarctic soils. Would they find enough of a soil community to make the dry valleys a worthwhile place to study? After all, they were looking for a place to study life as it works on Earth, not a surrogate for Mars.

In their first season in Antarctica and several to follow, Wall and Virginia and their research teams sampled hundreds of sites, wet and dry, in Taylor and several other dry valleys. What they found and what they didn't find were equally surprising. First, they were able to extract nematodes from nearly two-thirds of their samples—firm proof that most of the dry valley soils aren't sterile and that soil food webs exist. The average 2-pound bag of dry valley soil yielded 700 nematodes, and the liveliest soils they sampled yielded 4,000.[15]

Their second finding was that more than a third of their samples contained no nematodes at all—a phenomenon unique on earth.[16]

"This is probably the only place on earth where you can pick up a handful of soil and not find a nematode in it, then march several steps and pick up another handful and find nematodes," Virginia said. "In almost every other system, they're ubiquitous, and the numbers and the diversity overwhelm you even in trying to characterize one sample."

It's difficult to grasp how ubiquitous and varied nematodes are in the world beyond Antarctica. In a square yard of pasture soil, for instance, you could expect to find 10 *million* nematodes, along with similarly overwhelming numbers of microbes and myriad other soil organisms.[17] Pioneering nematode researcher Nathan Cobb wrote in 1914: "If all the matter in the universe except nematodes were swept away, our world would still be recognizable, . . . its mountains, hills, vales, rivers, lakes, and oceans represented by a film of nematodes."[18] In the dry valleys of Antarctica, that thin and patchy film would be composed of only three species of nematodes, all of them unique to this continent: *Scottnema lindsayae, Plectus antarcticus,* and *Eudorylaimus antarcticus.*

Worldwide, some 25,000 nematode species have been named, and more than 10,000 of these live in the soil or seabed or freshwater sediments. But the named species are just a fraction of the world's nematode diversity. Anywhere from an estimated half million to 100 million more nematode species are still awaiting discovery.[19] One soil nematode has become a celebrity of sorts: *Caenorhabditis elegans* is widely used as a "laboratory rat" and became the first multicellular organism to have its full complement of genes sequenced. A team of developmental biologists won a Nobel Prize in 2002 for work that revealed how the genes of *C. elegans* regulate the development of a single fertilized egg into an adult. Not surprisingly, however, the best-known nematodes are not the ubiquitous, microbe-eating decomposers but the small minority that parasitize us and our livestock and pets—intestinal roundworms, hookworms, and nematodes that cause elephantiasis and African river blindness—or those that cause substantial damage by feeding on crop plants.[20] Wall still works on nematode diseases of alfalfa and other crops as well as nematode roles in larger soil processes.

The Wormherders have been returning to the dry valleys for 15 years to learn which of the three worms live where, what conditions each requires, and what makes some of these barren-looking soils better neighborhoods than others.

Scottnema is by far the most abundant nematode and makes its living eating bacteria and yeast out in the dry, salty soils that dominate the valleys. In these arid reaches, the team usually finds *Scottnema* or nothing. Because of *Scottnema*'s abundance, nematode numbers are three times higher in the dry polygon surfaces than in moist habitats near streams and lakes.[21]

"*Scottnema* is king, he's just lovely," Wall said as she opened a greatly enlarged mug shot of the worm—only 1/25th of an inch (a millimeter) long—on her laptop one day. "Look at those probolae!" She pointed to wavy tentacle-like extensions encircling the head end. "And ruffles! Imagine if you had ruffles!" The wrinkly cuticle on the beast's "neck" resembles a stack of Elizabethan ruffs—or to be less charitable, the worm equivalent of a triple chin. *Scottnema* is indisputably a dandy among worms. The other two dry valley nematodes, by comparison, are plain as spaghetti noodles.

Byron Adams likes to ask his students to guess the function of *Scottnema*'s probolae. "You'll get answers like 'oh, they're feelers' or 'chicks dig the guys with the big long wavy things.' The truth is, we don't know."

Plectus, like *Scottnema*, eats bacteria, but it prefers living in ephemeral streams. *Eudorylaimus* is rarer than the other two and prefers damp places.[22] For years, *Eudorylaimus* was labeled an omnivore-predator and suspected of feeding on its fellow worms. Then last season, Adams photographed one of these transparent creatures with a gut full of algae, confirming instead that *Eudorylaimus* is a vegetarian. So far, in 20,000 soil samples examined over the years, the team has yet to see anything preying on a nematode. That's why Wall has dubbed them lions, kings of the food chain on this harsh plain.

As for the rest of the soil community, the Wormherders have found some tardigrades and rotifers in the wetter sites, and New Zealand researchers have found springtails and mites under surface rocks.[23] These findings confirm that the soils here, although certainly

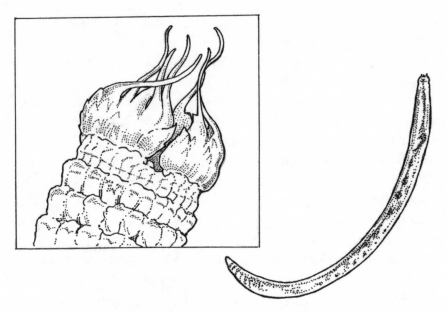

A microscopic close-up of the head end of the Antarctic nematode *Scottnema lindsayae* reveals tentacle-like probolae and a ruffled "neck."

not sterile, host the simplest food webs and lowest biological activity of any soils on earth. Biological activity refers to the daily business of breathing, moving, growing, eating, and being eaten that drives the process of rot and renewal that we call nutrient cycling. Microbial decomposition, or rot, for instance, proceeds so slowly that the dry valleys are littered with freeze-dried carcasses of seals like the one Scott's party saw, including some that died hundreds of years ago. The Wormherders themselves see many more carcasses, ones Scott couldn't have seen.

"One of the things that shocked me when we ran the first samples down here was that we'd see so many dead bodies in the soil," Wall had told me. She meant dead nematodes, tiny morsels that are quickly reduced to recyclable carbon, nitrogen, and nutrients in more amenable climates.

What scant biological activity there is in Antarctica drops to nothing when the sun and liquid water of summer disappear. Even in the summer, Wall pointed out, anywhere from 30 to 80 percent of the nematodes extracted from a dry valley soil sample will be coiled and dormant in a cryptobiotic state known as "anhydrobiosis"—literally, "life without water."[24]

Antoni van Leeuwenhoek, who devised the first microscope in the 17th century, was apparently the first person to witness a rotifer—he called it a "wheeled animalcule"—awakening from this dormant state. In the 18th and 19th centuries, the phenomenon of anhydrobiosis prompted a debate about whether the creatures were actually experiencing death and resurrection.[25] Anhydrobiosis is a drastic but reversible state triggered by dehydration. In the late 1970s, Wall's research showed for the first time that virtually all nematode species in hot deserts could undergo anhydrobiosis. When soil moisture levels drop below 2 percent, water films on soil particles dry up and desert nematodes begin jettisoning 99 percent of their body water. As the worms dry, the rings or annulus of their body draw closer together like a slinky toy recoiling, and they curl into a characteristic Cheerio shape. (Tardigrades collapse into a dried ball known as a "tun," and rotifers morph into tiny mushroom shapes during anhydrobiosis.) Internally, the worms begin producing an antifreeze solution such as trehalose or glycerol, which protects their membranes during desiccation. All detectable metabolic activity and respiration cease. Just add water, however—a dusting of melting snow or thawing of wet, frozen soil—and the worms begin to swell and uncoil; within 24 hours they are wiggling blindly around and turning their sensory powers to the search for food.[26]

Nematodes have been resurrected from this state from soil left on a shelf for 60 years or more, Wall pointed out, but no one knows the upper limit to such time travel. It confounds our sense of time and lifespan that a relatively brief life cycle—for dry valley nematodes, about 7 months in a warm, moist lab environment—can stretch for decades, perhaps centuries, of golden days interrupted by long, ageless sleeps through hard times. Nor is this talent for time travel unique to Antarctic or desert worms. Wall found nematodes coiled and dormant in agricultural soil, too, which helps explains why farmers cannot count on ridding their fields of plant parasites simply by leaving the fields fallow. Nematodes in the soil of an Iowa cornfield or an Amazonian forest or your garden can enter anhydrobiosis, although some may surrender and go dormant under far less water stress than natives of arid regions. In fact, a large fraction of the life forms in any soil community may be dormant at any given time, waiting for a growing root tip to shove past or a favored bit of detritus to fall into their sphere or environmental conditions to change at the soil-pore level. The business of life underground everywhere varies seasonally and minute to minute.

How did Antarctic nematodes learn to survive not just drying but freeze-drying? Adams envisions them wiggling beneath the feet of dinosaurs in a beech and conifer forest 200 million years ago when the chunk of the earth's crust that is now Antarctica formed the heart of the Gondwana supercontinent and enjoyed a climate more like that of Oregon: "I think these nematodes actually evolved here," he says. "I think that at one time Antarctica was extremely diverse, just like the northern and southern hemispheres. Then it moved down here where it got colder. But I think it got dry first and then cold. Most people down here think, 'isn't it amazing, these nematodes have evolved to live in cold temperatures.' But I think the opposite is true. I think what really happens is that they do what nematodes in deserts in California do. And it turns out that if you're able to dehydrate yourself in order to survive in a desert, you don't care what the temperature is. It's a key innovation that allows you to survive more than one type of extreme."

Adams and Wall are building a collection of DNA from nematode populations across Antarctica, from sites with different geologic histories and soils and varying degrees of isolation. The two hope to track the evolution of genes that affect the creatures' survival, their responses to the environment, and their contributions to ecological processes. With any luck, they may even come across ancient carcasses of nematodes or long-dormant worms locked in permafrost or glacial formations.

"It's like looking for dinosaur DNA," Wall says. From microscopic dinosaurs.

The two have been asking geologists bound for sites deep on the continent to bring back bags of soil from exposed patches of earth. Their prize acquisition so far has been a single *Scottnema* pulled from a sample from the Beardmore Glacier, which flows from the Transantarctics onto the Ross Ice Shelf at 83° south latitude. The genetic work done so far in Wall's and Adams' labs, however, has shown that *Scottnema* is essentially the same beast throughout the continent.[27]

"I was a little bit disappointed and also a little astonished, given the distances between the sites, to see that they were virtually identical genetically," Adams tells me one day out in the field. "How could this be? And the best I can come up with is they're either incredibly slowly evolving or there's this rampant dispersal." Dispersal of individuals would keep genes flowing between populations and overcome the genetic isolation that often allows new species to evolve.

A few days after that first field outing to F6, we are working in the worm farms at the south end of Lake Hoare. There are six of us this time, the original three plus Emma Broos and Johnson Nkem from Colorado State and Jeb Barrett from Dartmouth. Adams and Barrett—a veteran of multiple seasons on the ice—have been showing Broos and Nkem where to sample in one of the plots. It is another clear, brilliantly sunny day, but a brisk wind sweeps down the valley, chilling our bare hands as we scoop soil into bottles and bags.

The Wormherders have long believed that these winds blow dormant nematodes around the valleys like freeze-dried Cheerios. That's the reason for the dust traps at each field site. But Adams tells us he thinks wind dispersal of nematodes could occur on a much larger scale.

"The circumpolar winds could act just like a big toilet bowl, swirling them around the continent," he says, clearly peeved by the prospect because it bodes relatively uniform genetics.

Barrett sees it in another light: "It may not be an interesting result for an evolutionary biologist if they're all genetically identical, but for an ecologist it's great. It shows this is one tough little worm."

Tough because the same genes seem to equip *Scottnema* to live across a wide range of habitats.

"Isn't it an amazing beast," Wall says, passing by with two carboys of water. It is not a question.

Tough and amazing they may be, but how will nematodes the world over respond to shifts in their environment that are likely to accompany human-driven changes in atmosphere and climate? Since their first season on the ice, Wall and Virginia have been manipulating temperature, water, and carbon—the essential food stock of earthly life—in their worm farm plots in a search for answers. The cone-shaped chambers, for example, which now number more than 100 at three sites, act like miniature greenhouses, raising the temperature of the soil inside by 1–2° F. The researchers quickly learned that even this minor warming, far from being a boon to the worms, simply dries out the soil and knocks back nematode numbers.

Emma Broos and I follow Wall to one of the long-term plots and measure out portions of water or solutions of carbon-rich sucrose or mannitol—a sugar alcohol that worms would get eating cyanobacteria. Wall and Barrett then move along the rows sprinkling water or one of the sugar solutions on the gravelly soil of some sites, with or without chambers, and leaving others as is. After nearly a decade of these annual boosters, the water has done surprisingly little to benefit nematode populations. The sugar amendments, on the other hand, have indirectly boosted nematode numbers, except inside the chambers where increased drying caused by warmer temperatures cancelled the benefit. It took 8 years, but the sugar has supplied enough carbon to fuel a microbial population explosion that provides a feast for the worms.

In the green regions of the earth, soils are usually chock-full of carbon. Some of it is in highly edible forms that cycle quickly through the tissues of plants to the bodies of grazing animals to the decomposer microbes and out through the soil food web of consumers and predators until it eventually returns to the atmosphere. Most of the organic carbon in soils, however, is locked up long term in recalcitrant forms such as lignin and cellulose and the cakey, decay-resistant black humus that gardeners and farmers recognize as the sign of a rich and fertile soil. The capacity of soils to store carbon is particularly important at a time when human societies are pumping unprecedented amounts of carbon into the atmosphere from the burning of fossil fuels, enough carbon to alter the global climate. The soils of the earth harbor more than three times as much carbon as the atmosphere, and four times as much as the bodies of all living plants and animals.[28]

Little of that carbon, however, is here in the soils of the dry valleys. In a place with no trees, shrubs, or grasses, the task of assimilating carbon from the air through photosynthesis falls to microscopic green algae and chlorophyll-containing cyanobacteria. Consequently, these soils not only host the lowest biological activity but also the lowest organic carbon reserves of any soil on earth. Carbon, in fact, appears to be more important than moisture in defining good neighborhoods for nematodes and other soil life in the dry valleys.[29] Organic carbon stocks are richest in the centers of soil polygons where the most abundant *Scottnema* populations are found.[30]

One of the first questions Virginia and his students asked was, where did the meager carbon stocks in these soils come from? The long-standing assumption had been that the soils were passive beneficiaries of organic carbon blown out on the wind from eroding cryptoendolithic communities—the cyanobacteria living inside rock pores—and algal mats rimming today's streambeds and lakeshores. But it turned out to be a much older bounty. By reading the isotopic signatures of the carbon in valley soils, the researchers discovered that most of it is "legacy carbon" left over from the Pleistocene epoch (2 million–10,000 years ago) when a glacial lake known as Lake Washburn inundated Taylor Valley to a depth of 1,000 feet. Carbon had been pulled from the air by lush algal mats that rimmed the shorelines and carpeted the bottom of that ancient lake, and the carbon-rich mats got left behind as lake water receded.[31]

The realization that most of the carbon is an ancient legacy prompted another question: How much carbon are today's soil organisms using? All living things, from plants to nematodes to humans, "burn" carbohydrates and other carbon-based molecules to extract energy to fuel their life processes. This metabolic activity generates water and carbon dioxide as wastes, and organisms respire or exhale the CO_2 back to the air.

"The simplest way to 'black box' the entire soil ecosystem is just to measure the exhaust pipe, which is the CO_2 being exhaled," Virginia said. With no plant respiration to drown out the breathing of the microbes and nematodes, the soil exhalation is so faint that it took his group 2 years to get the right instrumentation to measure it. When they did, they found the balance sheet didn't add up.

"Just based on the low levels of respiration we measure, the soil communities would have exhausted this legacy carbon in a matter of decades," Barrett recounted. There has to be some ongoing production by algae in the soil and some carbon blown on the wind from contemporary streams, lakes, and ponds.[32] But is it enough to fuel life in the valley today? Each season the team had been measuring chlorophyll in the soil—a proxy for carbon production through photosynthesis—but the numbers were not high enough to balance the budget. By the late 1990s, it looked as though the soil community might be eating up more than it produced, drawing down its legacy and spiraling downhill.

The idea that these tough worms and their compatriots were in trouble was soon reinforced when the various research teams studying the dry valleys began to compare notes.

In 1993, soil researchers teamed up with scientists from a number of disciplines—microbiology, geochemistry, hydrology, and glaciology—who had long been studying the dry valley lakes, streams, and glaciers. They developed a tight-knit collaboration that won funding as the McMurdo Dry Valleys Long Term Ecological Research program, one of two-dozen "LTER" programs supported by the National Science Foundation. Some 30 researchers, from senior scientists to students, converge on Taylor Valley each austral summer, sharing the four seasonal field camps and working side by side. After the first 6 years, when the LTER scientists got together back in the United States to pool their findings, they discovered to their surprise that the situation looked grim for life in the dry valleys. Here's the gist of it:

Contrary to the warming trend over most of the globe, and even on the Antarctic Peninsula, part of the continent has been growing colder recently. The dry valleys cooled 1.3° F per decade from 1986 to 2000, with cooling even more pronounced in the summer. The cooling spurred a rapid cascade of consequences in the valley's ecosystems. Glaciers melted less, stream flow thus decreased and lake levels dropped, and the permanent ice covering on the lakes grew thicker. Soil moisture dropped, too, by more than one-third. Algal production in the lakes declined by 6–9 percent a year. In Wall and Virginia's worm farms, nematode and tardigrade numbers dropped more than 10 percent each year. Between 1993 and 1998, their populations shrank by half.[33]

The odd pattern of heating and cooling across the Antarctic may be triggered by complex oscillations in atmospheric pressure that alternately speed and slow the fierce circumpolar winds. Some atmospheric scientists argue that human-caused ozone depletion—specifically, the ozone hole that forms high in the stratosphere over Antarctica each spring—may be at least partly to blame. Without the ozone layer to absorb the sun's energy, the stratosphere cools each spring, intensifying the westerly winds that circle high above the continent. At certain seasons, these stratospheric winds are believed to influence pressure and thus winds in the lower atmosphere.[34] Whatever the cause of the cooling, life here seemed to be spiraling downward faster than anyone imagined.

Even as the group's findings were being published in a scientific journal, however, the outlook changed dramatically. Near the end of December 2001, what passes locally for a heat wave descended on Taylor Valley, raising temperatures above freezing for almost a month. Temperatures rose as high as 50° F; meltwater pooled on the tops of glaciers and poured down in waterfalls; streams overflowed their banks, topping previous flood levels by two- to threefold and sometimes cutting new channels; stream water tumbled into lakes, restoring in a matter of weeks lake levels that had been dropping for a decade; soils near waterways were soaked. By the end of January 2002, the wet soils froze, retaining their moisture when they thawed the following season.

That next season, the Wormherders saw a mixed response from their charges: Water-loving *Eudorylaimus* quadrupled its population and responded with a quick population boomlet when given a carbon feeding. *Scottnema*, however, seemed to get no benefit from the extra moisture and has continued the decline begun in the cold years.

Perhaps the most significant revelation involved carbon production.

"After the wet year, you'd go out there and on the surfaces of the light-colored rock you'd find layers of green slime," Barrett recalled. "If you'd take this algae back and incubate it, you'd get high rates of photosynthesis in growth chambers." Chlorophyll concentrations in the soil spiked, indicating a pulse of relatively luxuriant carbon production by soil algae.

The wet year gave the team a new way of looking at the soil ecosystem here, Wall explained. What looked like an inexorable decline in the 1990s was perhaps just a downswing in a cycle that perpetually runs close to the edge, only to be reset sporadically by a burst of warmth and wetness, a pulse of productivity. Perhaps life in these valleys has long seesawed between doom and boom.

"The legacy of that pulse can persist on timescales you wouldn't see or think of in other ecosystems," Virginia added. "The green slime is our little redwood forest. That stuff may persist on the same timescales as ancient trees, locking up and holding carbon for thousands of years. This system cycles on a timescale that most ecologists don't work on. That's one thing that really changed for me coming down here. You have to think in geological time, glacial cycles. Here, the timescales of the ecology and the geology become almost the same."

The 1990s cooling followed by the warm, wet year reinforced the perception that life exists close to the edge here, exquisitely sensitive to small changes.

"We've seen that a half a degree change in temperature one way or the other can have a tremendous impact on lake levels, soil moisture, the nematode communities," Wall pointed out. "Even when a cloud comes over, boom, the soil temperature can drop and the whole system shuts off. So imagine what will happen if we really get a significant climate change down here."

The drastic drop in *Scottnema* populations has given the team an unexpected opportunity to see whether the changing fortunes of this dominant consumer will have a detectable impact on the flow of energy and nutrients through the dry valley food web. Field tests using isotope-labeled carbon have already shown that *Scottnema* is responsible for a disproportionate share of the CO_2 exhalation from these soils. Indications are that the identity and abundance of the nematodes at a site have more ecological significance than the sheer diversity of species. In the dry valleys, the loss of *Scottnema* would likely have a much greater impact on nutrient cycling than the loss of either of the other two nematode species.[35]

Climate change isn't the only disturbance that could put these ideas to the test. Wall and the other Wormherders also worry about how the soil community will respond to increasing physical disturbance by scientists and tourists in the dry valleys—a place she refers to, only half in jest, as Nematode National Park. Already, the first cruise ship of the season has pulled into McMurdo Sound, offering its passengers helicopter tours to the legendary Mars-like dry valleys.

The fortunes of *Scottnema* may have little import beyond these valleys. Yet the very sensitivity of this worm and its companions may help ecologists answer questions that are key to reducing the human footprint in more complex ecosystems where the integrity of soil communities is vital to our own lives and well-being.

III

Of Ferns, Bears, and Slime Molds

Newfound Gap Road twists for 26 breathtaking miles across the mountainous heart of Great Smoky Mountains National Park, climbing from the northern gateway at Gatlinburg, Tennessee, across a 5,000-foot divide to Newfound Gap, where it intersects the Appalachian Trail, and then on to join the Blue Ridge Parkway in North Carolina. This is the most visited national park in the United States, and most of the 10 million people who flock here each year will drive up at least as far as Newfound Gap to enjoy the ridgetop views and perhaps venture out a ways on the Appalachian Trail.

It is a sunny Saturday morning in early August when I arrive at Newfound Gap with park service biologist Keith Langdon. The parking lot is already beginning to fill, and people are piling out of cars with daypacks and cameras. This is not going to be an ideal day for scenic photographs, however. Despite several days of cleansing rains, bright haze shrouds the green slopes nearby and obscures the ancient mountains beyond.

The Cherokee Indians called these mountains the "place of blue smoke" because of the forest-generated haze that frequently clings to the hollows. Today's opaque white skies, however, are created not by

trees but by power plant and automobile exhaust generated throughout the eastern half of the United States. The pall of airborne sulfates,
nitrates, and ground-level ozone is at its worst on summer days like
this.[1] The National Parks Conservation Association has consistently
included the Great Smoky Mountains on its list of America's 10 most
endangered national parks, and in 2002 declared it the most polluted
national park in the nation.[2] Even more worrisome than the loss of
mountain views are the unseen effects of the park's deteriorating air.

As we climb out of Langdon's car and begin collecting our hiking gear, two young men nearby are setting up a card table under an
awning. Their sign reads "Western Carolina University Hiker Health
Study." Langdon encourages me to go over and blow into a device
called a spirometer that will test my lung capacity before and after our
planned hike. Up here on the ridges of the park, I later learn, we are
breathing ozone concentrations that average as much as twice what
we would experience in urban Knoxville or Atlanta.[3]

Human visitors aren't the only potential victims of air pollution,
and that is part of the reason for our hike today. Some 90 plant species
in the park show leaf damage characteristic of ozone injury, and ozone
is suspected of slowing the growth of sensitive plants such as tuliptree
and black cherry. The park also suffers some of the highest acid precipitation levels in the nation, with sulfur and nitrogen pollutants
falling in the form of dry particles as well as rain and shrouding
mountaintop spruce-fir forests in fog as acidic as vinegar. Already,
some high elevation streams suffer excessive nitrate levels and some
forest soils are saturated with nitrogen, a condition that causes leaching of vital soil nutrients such as calcium and makes potentially toxic
aluminum more available to trees.[4]

Unfortunately, air pollution is not the only threat to plant and
animal life in the park. Invasive species such as feral pigs wallow and
root, destroying plants and soil communities; imported aphid-like insects known as balsam woolly adelgids and hemlock woolly adelgids
literally suck the life out of Fraser firs and hemlock trees. There are
also pressures from growing numbers of visitors, millions drawn here
not just by the rugged mountains but also by the maze of theme parks,

bungee jumps, petting zoos, factory outlet malls, golf courses, wedding chapels, and resort home developments proliferating along the park boundaries and isolating it from nearby natural areas.

Langdon and other park personnel can spot woolly masses of adelgids attacking pollution-damaged trees, black bears hit by cars, and ferns and rare orchids plowed up by foraging pigs. But it's hard to monitor the fate of lesser known, highly diverse groups in the park that fall under their stewardship. Most of that life, particularly soil life such as mushrooms, snails and slugs, springtails, and slime molds, has only been sporadically collected or cataloged, and the diversity of most small creatures in the park remains completely unknown, much as it does in the rest of the world. Indeed, there is no patch of soil or sediment on earth where we know the identity of every creature. Yet we know that many of the mushrooms in the park spring from soil fungi that form nurturing partnerships with tree roots, and other creatures from springtails to slime molds are part of the great web of decomposers that maintain the fertility of these soils. How would Langdon and his colleagues know if these creatures were in trouble or understand how their loss might ripple through the community?

Conserving soil communities is surely a worthwhile goal in its own right, and part of the mission of the park. But knowing what lives in soils and sediments, where it lives, and what it does will also provide a stronger foundation for maintaining the health of the lands and waters we rely on for food production, timber, drinking water, and other goods and services. Soil life can also be viewed as a valuable and largely untapped resource to be explored for new antibiotics and pharmaceuticals, industrial enzymes, novel genes useful to genetic engineers, and versatile microbes that can be harnessed to counter pests and pathogens or devour and detoxify hazardous wastes. Even more vital is the role soil creatures play in sustaining the ecological processes that keep nutrients cycling, soil renewed, plants growing, water fit to drink, and soilborne pests and diseases in check.

With this in mind, the Great Smoky Mountains in 1998 became the first protected area in the world to launch an ambitious effort to inventory literally every species in the park. Interested scientists and educators formed a nonprofit organization called Discover Life in America, Inc. to direct a 10- to 15-year effort called an All Taxa Biodiversity Inventory (ATBI) under a cooperative agreement with the park.[5] (Taxa are groups of related organisms.) The effort is funded by private donations and grants. Some 200 participating scientists and hundreds more lay volunteers join in organized sampling forays that involve literally beating the bushes for spiders; searching deep caves for millipedes, daddy longlegs, and rare amphipods; picking fleas, lice, and other parasites off netted birds; trapping bats, shrews, and voles; dissecting biting flies to search their innards for symbiotic viruses, bacteria, protozoa, and parasites; pawing through forest leaf litter for tiny fungus beetles; bagging up thousands of soil samples to be screened for unseen life forms such as nematodes and springtails; and searching under rotten logs for signs of leeches reputed to feed on earthworms. Taxonomists team up with volunteers several times a year for concentrated "bio-blitzes"—a Millipede March, Protozoan Pursuit, Snail Search, Bat Blitz, Lepidoptera Quest, Beetle Blitz, Fern Foray, Fungi Foray, or in today's case, a High Country Quest.

This morning I'm tagging along with Langdon, summer intern Donelle, and volunteer Dick—an engineer and amateur photographer from Knoxville—for a 5-mile round-trip on the Appalachian Trail to sift soil, shake trees, and search through leaf litter. We are one of a dozen teams collecting samples from high elevation habitats in the park this weekend. Another is an international team of taxonomists who are searching the high country for slime molds, odd creatures that inhabit tree bark, litter, and soil and make their living eating bacteria.

The students from the Hiker Health Study aren't finished setting up their equipment, so we gather up sample bags, vials, nets, and other gear and start for the trailhead without testing our lungs. Where the pavement ends, we pass a portable ozone monitoring station mounted on a metal railing, silently documenting what we and the creatures we seek will be breathing today.

Walking east, we quickly come to a "beech gap," one of the most rare and endangered plant communities in the region. Half of the beech trees in the park, Langdon tells us, are dead or dying thanks to an alien scale insect that carries a deadly fungus that infects the bark. Likewise, half of the flowering dogwood has succumbed to another alien fungus.

We walk on through narrow avenues of rhododendrons, the white-flowering ones in full bloom and the dark pink Catawbas just fading. Beneath them, amid the damp, mossy rocks, are ferns and numerous tiny mushrooms. Some of these mushrooms are the fruiting bodies of soil fungi that form mycorrhizae (pronounced micah-rizah)—literally "fungus roots." These fungi enter into symbiotic partnerships with the roots of most of the world's plants, drawing on a plant's sugar stocks for nourishment and in turn, sending out thread-like hyphae to serve as root extensions, absorbing water, phosphorus, and other nutrients the plant needs and offering some protection from pests and pathogens in the soil. The mycorrhizal fungi we are looking at here are called ectomycorrhizal fungi because they grow around plant roots like a sheath. They also produce mushrooms.

"I've been to Costa Rica a couple of times and I was shocked that I would go all day hiking in the tropical rain forest and I'd find maybe one mushroom," Langdon recounts as we move out onto a forested knife ridge, its sharp slopes obscured by lush greenery. After 19 years in the Smokies, he radiates an unabashed pride in all its wonders, from mushrooms to salamanders. "Here, you can stand in one place and look around and see 20 different kinds of mushrooms all growing out of the duff," the spongy layer of decaying leaf litter and organic debris on the forest floor. Most tropical trees partner with another type of fungi—arbuscular mycorrhizal fungi, whose hyphae actually penetrate and grow into the root cells—that do not form mushrooms.

Langdon's mention of Costa Rica reminds me that the concept for an ATBI started as the brainchild of University of Pennsylvania tropical ecologist Dan Janzen, who beginning in the late 1980s set about laying the groundwork for the world's first all-taxa inventory in the rain forests of Costa Rica's Guanacaste Conservation Area.[6] Costa Rican officials, however, opted in 1996 to survey a limited number of key groups such as plants and fungi at five conservation areas instead of inventorying every species at one area. Not long afterward, Janzen and other ecologists began talking with the U.S. National Park Service about the possibility of launching an ATBI in the Great Smoky Mountains.[7] Langdon and his colleagues arranged a meeting for December 1997 in Gatlinburg to try to gauge interest in the idea. More than 100 ecologists, taxonomists, educators, and park administrators showed up to begin planning the survey.[8]

Biologists expect that this half-million-acre park in the Smokies harbors only one-quarter the plant and animal diversity of a tropical rain forest. Yet that diversity represents the richest array of plant and animal life in temperate North America. For millions of years, the Smokies have flourished as an ice-free refuge for northern species driven south during repeated glaciations, and the mountains still offer a wide array of climate and life zones. Driving up to Newfound Gap this morning, we had traveled through five botanical life zones, from lowland cove hardwood to high-elevation spruce-fir forests. The park's status as a hotspot for temperate biodiversity has earned it United Nations recognition as a World Heritage Site and an International Biosphere Reserve.

What's more, current comparisons of the richness of tropical and temperate diversity may be misleading. As I heard repeatedly from soil ecologists, the notion that life reaches peak diversity in the tropics comes only from observations aboveground. There's little evidence that life underground, the most diverse life on the planet, follows the same rules. The diversity of termites does increase as you move from temperate regions toward the tropics, but the diversity of nematodes, earthworms, protozoa, and fungi apparently does not.[9] Neither does the diversity of slime molds, according to Steve Stephenson of the University of Arkansas, who leads the slime mold "taxonomic working group" or TWIG now assembled in the park. "Some of us think that the total number of slime molds here in the Smokies will rival that of any place on Planet Earth," he had told me. Why they are distributed across the earth as they are, Stephenson can't answer. But he is quick to respond when I ask what their virtues are: "The slime molds are one component of the earth ecosystem—obviously, to most people, a rather insignificant component. But they feed upon bacteria and release nutrients back into the biosphere. If we didn't have them, bacteria would tie up those nutrients, and I suspect the system would be very different."

When the ATBI began, biologists came up with an ad hoc estimate that there could be 100,000 species in the park, not counting microbes, although only about 9,800 species had been reported—and only a tiny fraction of those known species were soil and sediment creatures.[10] By the end of 2004, according to Discover Life in America's running scoreboard, participating scientists had identified more than 3,300 previously overlooked species in the park. Among these are more than 500 species new to science—species never before collected or named anywhere on the earth—and many of those live on or in the soil. They include 92 bacteria, 3 fungi, 67 algae, 9 lichens, 20 slime molds, 8 tardigrades, 36 springtails, 4 earthworms, and 3 snails.[11] Thousands more specimens yet to be identified float in vials already stored on shelves and in freezers in the cluttered basement of a cabin at the park's natural resources office or in the labs of participating specialists around the world. Identifying species, while fundamental, is not the ultimate goal of the inventory. Park personnel want to know what community or habitat a species favors, the size of its populations, what season it appears or hibernates or blooms, what it does for a living, and how it interacts with things it eats, feeds, pollinates, partners with, or parasitizes.

"One of our goals, of course, is to really strengthen our monitoring program," Chuck Parker, a U.S. Geological Survey biologist stationed in the park, had told me. "We're convinced this requires having a much more thorough understanding of what's here before we make decisions about what it is that should be monitored." The idea is to find "canaries" or sentinel species that are highly sensitive and respond early to changes in the environment and that can also be censused regularly with minimal effort and cost. These sentinels could alert park staff to insidious changes they might not otherwise recognize until irreversible damage has been done.

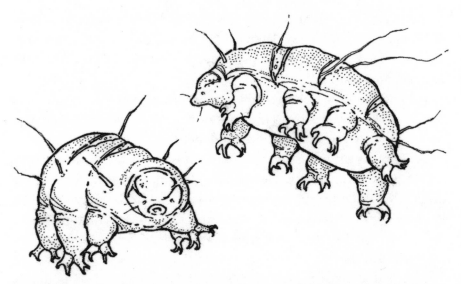

New species of tardigrades are among the tiny wildlife discovered in the Great Smoky Mountains. Tardigrades—about the size of the dot over this "i"—are called "water bears" because they live in water films on lichens, mosses, or sand grains.

Our trail drops below the ridge with the north-facing slope rising steeply on our right. Langdon stops and proclaims this a good place to begin sampling. Dick and Donelle set to work, one whapping the branches of a cranberry bush with a stick and the other holding up a "beat sheet" to catch whatever bugs and beetles rain down. Then Donelle uses a rubber tubing and pipe apparatus known as a "pooter" to suck the catch into a jar.

Langdon clambers upslope through the ferns and shrubs a short distance and begins pushing aside the moss and duff, filling zip-top bags with moist, humus-rich black soil. The gallon bags will go back to park headquarters near Gatlinburg where the soil will be placed in funnels with lights suspended over them to "motivate" tiny organisms inside to crawl downward and fall into waiting flasks. Smaller bags of soil will be delivered to the slime mold team to be cultured in the lab for unseen amoeba-like cellular slime molds and other tiny slime molds known as protostelids. Like nematodes, cellular slime molds have one celebrity among them: a creature known as *Dictyostelium discoideum*, which has long been a "white rat" in biology labs. Scientists know its genes and enzymes well, but its life in the wild remains unexplored.

Langdon says we're going to sift another bag of soil right now for visible beasts. He tosses the bags down to me and works his way back to the trail to show me how to label them.

He reads off two strings of numbers, coordinates taken from a global positioning device. He adds the date and "Appalachian Trail, east of Newfound Gap, approximately 1 kilometer." "That's so even if someone transposes a number and this grid coordinate turns up in Botswana, we'll know about where it came from and be able to correct it," he adds.

Langdon takes a small-mesh sieve, pours a measured amount of soil and litter on it, and shakes it vigorously while I hold a pan underneath, catching the soil crumbs.

"Now we're going to look for everything that moves in here," he tells me, instructing me to sit with the tray of crumbs in my lap. "A lot of things won't move for the first couple minutes, so just be ready." I put the beige rubber tubing of a pooter in my mouth and hold the metal pipe end poised over the tray, waiting.

"This is something over here, really tiny, a spider," Langdon points. I move the pipe end over the crawling speck and suck. Success—it's in the jar, but so is a fair amount of dirt. I bag a tiny beetle, then several springtails. Over the next 15 minutes, I capture a few dozen more moving specks. Finally, nothing else on the tray seems to be stirring and Langdon holds a specimen vial while I tap the contents of my jar into it, ending with one reluctant springtail. I add alcohol and slip in a pencil-written label with the collection date and place and my name. Now I have an official sample in the inventory.

At Langdon's direction, we all begin a search for snails. "Snails are going to be at the base of trees," he says. "They'll be under the moss, under rocks, and it helps sometimes if you take your fingers and just claw gently through the leaf litter and see if anything comes out."

I ask what my search image will be, what color.

"They'll be horn-colored. But the thing is, there's so much acid falling at the upper elevations in these mountains per year that the mollusk specialists are telling us that there's not the snails there used to be, either in number or type. And I don't think we're going to find much here, frankly."

We do find a few, then gather our gear and move on up the trail. Within an hour, rain is sheeting down on us. I've left my raingear behind and soon I'm drenched. The others are better prepared. The trail becomes an ankle-deep stream as we hurry on to the backcountry shelter at Icewater Spring. The rain is warm and my soggy condition not uncomfortable, but I wonder what pollutants are being washed from the sky with this downpour.

If snails and trees already suffer from the acid levels, they are unlikely to be suffering alone. Many studies show that acidification of forest soils causes shifts and losses in plant and soil communities.[12] Across central Europe, for example, where acid rain began causing noticeable damage to spruce and fir forests at least three decades ago, researchers have documented declines in the abundance and diversity of mycorrhizal fungi. Mushroom fanciers know that boletes, chanterelles, truffles, and many other prized forest delicacies are the fruiting bodies of mycorrhizae, and these have been growing scarcer since the 1970s in the forests of Europe.[13] I can't help wondering about the health of the rich assortment of boletes and other fungi here that Langdon points to with such pride.

Late in the day we return to park headquarters at Sugarlands Visitor Center. Other teams are returning from their outings, too, and all of us are unloading samples and assorted collecting equipment onto the tables. Jeanie Hilten, administrative officer of Discover Life in America, appears with pizzas for everyone. Hilten has been out leading another team collecting salamanders, and later she and some volunteers will put today's soil samples into funnels to begin extracting tiny soil animals. Entomologist Ernie Bernard from the University of Tennessee at Knoxville will pick up the material in a few days, sort it, and ship some of the creatures off to other specialists. He himself will identify the springtails and the protura—an "oddball group" of six-legged, wingless creatures that are similar to springtails but have lost their antennae and so walk about on four legs, waving the front two in place of antennae.

"Protura are amazingly original creatures," Bernard had told me earlier when he dropped off the funnels. "There's nothing like them at all." Springtails, too—formally, *Collembola*—display an originality captured in their common name. It refers to the spring-like appendage called a furca folded under their hind end that allows many to propel themselves forward like pole-vaulters. They are the most abundant six-legged creatures on earth, more abundant than ants and termites and extremely ancient. Springtails feed on decomposer fungi; apparently so do protura, although little is known about their occupation. Both springtails and protura share a recent distinction: molecular work has led many specialists to conclude that they are more closely related to crustaceans such as shrimp and copepods than to insects.[14]

Ernie Bernard seemed to relish recounting these idiosyncrasies, and I commented on it.

"Yes, one thing taxonomists want is they want their creatures to be weird," he confirmed enthusiastically. "You don't want to be working on just brown beetles!"

A taxonomist's affections can be quite focused, too. When Bernard comes across primitive arthropods known as pauropods— smaller and lesser known cousins of millipedes and centipedes—he sends them off to the world's only pauropod specialist, a retired high school teacher in Sweden named Ulf Scheller. Scheller, who has also traveled to the Smokies to sample, calls pauropods "delightful and charming creatures." He is talking about barely visible (up to 1/16-inch long) beasts that go about under rocks and rotting wood or in the rich surface layers of the soil, nibbling mold or sucking the juices from fungal hyphae or root hairs.[15]

Unfortunately, specialists with both the knowledge and the enthusiasm to identify living things, especially soil creatures, have been growing harder to find.[16] Taxonomy and systematics have been unfashionable for decades and short of funds, computer database technology, and new recruits, despite the urgent need for information about the earth's biodiversity.[17] Harvard biologist Edward O. Wilson, who has written eloquently about the growing extinction crisis and championed a revival of taxonomy, says, "To describe and classify all of the surviving species of the world deserves to be one of the great scientific goals of the new century."[18]

Interest in and funding for taxonomy are on the rise in a few places, however, including other U.S. national parks. Inspired by the project in the Smokies, Point Reyes National Seashore in California launched a 5-year ATBI in 2003 to survey the biodiversity of adjacent Tomales Bay. Since then, a half dozen other national parks as well as the state parks of Tennessee have started planning ATBIs. And there are numerous other types of surveys, inventories, and biodiversity assessments under way for key groups of organisms in protected areas around the world, although many such efforts ignore life in the soil.

Perhaps the most ambitious survey effort to date was launched in 2003 when the U.S. National Science Foundation teamed up with the ALL Species Foundation[19] to launch a new global biodiversity tallying strategy called Planetary Biodiversity Inventories. Under this initiative, international teams will each focus on censusing a single group of organisms worldwide rather than all organisms in a single place. The first set of awards, totaling $14 million, is designed to assess the feasibility of completing such global surveys "within reasonable time frames."[20] The four groups of organisms targeted in the first round include all the world's catfishes, plants in the genus *Solanum* (which includes potatoes and tomatoes), plant-feeding insects in the family *Miridae*, and the *Eumycetozoa*—slime molds.

"Most of the people in this room will be involved in some aspect of the world inventory," Steve Stephenson tells me as the slime mold TWIG members unpack their day's samples. He will direct the new global effort as well as continuing the ATBI in the Smokies and mapping the distribution of various slime mold species to the plant communities in which they occur throughout the park. The crowd in the training room today includes experts from England, Lithuania, India,

Costa Rica, and the Ukraine as well as the United States, and this is only a portion of the slime mold TWIG team. With the majority of the world's experts already working together here, the team was better prepared than most taxa groups to go after the global project, Stephenson says. In addition, the anticipated diversity of slime molds seems manageable—perhaps 1,200–1,300 species worldwide—although in individual numbers these creatures probably outnumber springtails. Slime molds fall into three groups: myxomycetes, cellular slime molds, and protostelids. "Myxos" are 10 times more diverse than either of the other two groups—875 species are known so far—and they have been found everywhere from hot deserts to the Arctic and the Antarctic Peninsula. Myxos live their individual lives as amoeba-like single cells known as plasmodia, foraging for bacteria to engulf and digest. Then at some yet-unknown signal, groups of myxos assemble and team up to form a slug, which sends up tiny mushroom-like fruiting bodies. That is truly enough weirdness to capture the allegiance of a global team of taxonomists.

Inventories can set the stage not only for conserving and harnessing the resources of soil life, but also for exploring questions about underground ecology. For instance, soil ecologists want to know: What makes some sites richer in soil species than others? Does a higher diversity of plants aboveground encourage greater diversity of soil creatures below? In turn, does the diversity of life underground affect the diversity of plants above? Would it matter to the firs and ferns and bears of the Smokies or to human societies if soil systems were reduced to skeleton crews? Who are the most important players underground, and what are we doing that threatens or compromises them?

Similar questions have been hot topics for more than a decade for ecologists working aboveground. For obvious reasons, however, experimenting with uncensused millions of largely invisible species in an opaque medium makes fieldwork with soil organisms quite daunting. Even sampling relatively well-known groups such as springtails or nematodes can be overwhelming, except, as we saw, in the Antarctic Dry Valleys, where diversity is reduced to single digits. Rather than dealing with soil creatures as individual species, scientists usually lump them into broad functional groups according to their occupations —grazing on fungi, shredding leaves, munching on plants, or preying on other soil animals, for instance; or converting ammonia to nitrite, or nitrite to nitrate, both key steps in the recycling of nitrogen; or churning and aerating the soil, the work of "bioturbators" or "ecosystem engineers" such as earthworms on land and polychaetes in sediments. Assigning soil creatures to functional groups gives us only a rough cut, however, since we don't yet know how most of them go about making a living and to what larger functions their activities contribute.[21] Nevertheless, ecologists are picking up the pace of research on life underground.

One of the questions that has puzzled soil ecologists is how such a rich array of species can coexist in the soil. One piece of the answer lies in the tremendous diversity of habitats available at various scales, from tiny soil pores to clumps or aggregates of soil particles to larger patches created by the engineering work of ants or earthworms or the roots of plants, on up to landscape-level variations created by different soil and vegetation types and even human activities such as farming. Likewise, food resources such as plant litter, root secretions, dung, carcasses, and prey items are patchily distributed throughout the soil, and the timing of their availability is highly variable. Competition, a major factor limiting how many species can pack into aboveground communities, is probably not as important belowground because potential competitors are often physically isolated by their limited mobility and the complexity of soil habitats. Nematodes and protozoa both feed on bacteria, yet they probably compete very little, for instance, because protozoa can pursue their prey into tiny soil pores where larger nematodes cannot go. Besides, as we've seen, only a fraction of the organisms in a community are active at any given time. Microbes and soil animals alike spend much of their lives in dormant states.[22]

I realized at some point while researching this book that describing the soil as "teeming" with life and repeating the superlatives— "one cup of soil may hold as many bacteria as there are people on Earth"[23]—while true enough, had given me a misleading vision of that unseen world. I pictured an urban scene of elbow-to-elbow crowds jostling for existence. One of the people who helped alter that image was microbiologist George Kowalchuk at the Netherlands Institute of Ecology, Center for Terrestrial Ecology in Heteren.

"If you just think of it as one big soil system teeming with life where everything's competing with each other, then obviously you wouldn't have all these species," he explained. "You'd have a couple that would out-compete the rest and that would be it. You think of the soil as really densely colonized, but it's probably just little islands of colonization with large gaps between. Things like competition might not actually be relevant because you have to interact with your neighbors to compete with them. And if the next bacterial cell is, from your own perspective, kilometers away, then it's not going to compete with you for food or space. The soil is a heterogeneous place where you have little patches of different activity. You have these little microbial colonies, little towns, and each is probably rarely in contact with the next one because their inhabitants move slowly or not at all."

"The vast majority of microbial cells in the soil are dormant, waiting for their opportunity," Kowalchuk continued. "The typical soil microbe is playing a waiting game. If the wait is too long, they'll eventually die, but they can wait really a long time, perhaps years. Then when a root comes along or a drop of water or whatever, the cell blooms into a colony. And wherever you have a bacterial bloom, you'll certainly have a bloom of predators coming to eat those. Then once the plant or source of riches is gone, the colony probably dies out quite quickly and the few remaining microbes go back to waiting."

Breaking dormancy is often a gamble, as microbiologist David Hopkins of the University of Stirling in the United Kingdom explained to me. "Bacteria have a huge metabolic repertoire, but they don't keep all of the enzymes synthesized all the time," Hopkins said. "That's a waste of resources, particularly if they're in a dormant state. An organism may be sitting there, or a little colony, and they're all half-starved, they've all gone into physiological shutdown. And then immediately next to them an earthworm cast is deposited or a great big cowpat lands on top of them, and carbon starts diffusing into the soil. The organisms that respond and switch on at that first signal will be all tooled up and ready to go when the rest of the carbon arrives. But they are basically speculating, because they expend more carbon in preparing their metabolism than they actually get from that first little burst of carbon they're responding to."

Some of the same factors that allow soils in general to support high levels of biodiversity—that is, a wide diversity of habitats and food resources—can also create hotspots that are richer in species than others. Plants and soil animals such as earthworms both influence the physical character and the resources in the soil around them, and a fundamental link exists between plants, which produce carbon compounds through photosynthesis, and the soil community that decomposes those organic compounds. That in part is what has led soil ecologists to ask whether the number of plant species aboveground affects species numbers in the soil and vice versa.

The few studies conducted so far have shown no consistent link based on sheer numbers of species above or below the ground.[24] The relationship between plant species richness and the soil community may vary from one group of soil organisms to another, and even within groups—say, between nematodes that feed on specific plants and nematodes that graze on microbes. A project led by Colorado State University researchers, for instance, found no consistent trend in nematode diversity or soil properties between field plots planted with one versus two different species of prairie grasses. The results suggested that the traits of individual plants are more important to denizens of the soil than the diversity of plant species.[25] Another team from the Netherlands Institute of Ecology conducted a 3-year field experiment with mixtures of up to 16 plant species and found that both the number of plant species and the identity of the plants in the mix affected the diversity of nematodes in the soil food web. But the number of species apparently mattered only because it supplied an array of plant types that vary and complement one another in the quality of food resources they provide to the soil community. Nematode diversity varied more from one kind of plant to another than between different levels of plant species richness.[26]

Many other studies confirm that the identity of the plant species and the makeup of the plant community aboveground can influence both the diversity and abundance of certain soil organisms for better or worse.[27] This should not be surprising: plants vary in traits that critically affect the soil community, such as the nutritional quality and amount of litter they produce, the timing and amount of their root exudates, the depth of their roots, and how much they shade and cool the soil.

Plants also may "choose" which components of the soil community to support. Kowalchuk and his colleagues found clear differences in the makeup of the microbial communities in the rooting zone of two plants, hound's tongue and spear thistle, growing in the same experimental field. Molecular techniques also revealed that the diversity of microbes around the root zone or rhizosphere of each plant species was actually lower than microbial diversity in the bulk soil nearby. That suggests that each type of plant encourages a different subset of the local microbes within its sphere of influence, yet exerts little or no effect on microbial communities beyond the root zone.[28]

Kowalchuk expects that molecular techniques will allow researchers to further fine-tune the broad questions they have been asking about the links between diversity above- and belowground. For instance, some microbes such as root symbionts and disease agents that interact closely with plants should be sensitive to changes in the plant community. Other microbes "couldn't care less what type of plants are there," he believes. "They're only interested in what the total nitrogen load or phosphorus load is, or the total pH, things like this." Already he has found that ammonia-oxidizing bacteria—key workers on the assembly line of the nitrogen cycle—are oblivious to the diversity or identity of plants above them.[29] In contrast, the diversity of mycorrhizal fungi that live in intimate contact with roots sometimes declines along with the diversity of plants.

The flip side of this research is finding out whether changes in the soil community affect the plant community. Clearly, organisms such as earthworms and termites that alter the structure of the soil and organisms closely associated with plant roots such as disease agents, root-feeders, nitrogen-fixing bacteria, and mycorrhizal fungi can influence the makeup of the plant community.[30] And another telling experiment at the Netherlands Institute of Ecology, which we'll return to in a later chapter, recently showed that the makeup of the soil animal community, including nematodes, mites, and beetle larvae, strongly affects the composition of natural plant communities.[31] Recent studies by Heikki Setala of the University of Helsinki and his colleagues have shown that some species of soil animals in the decomposer food web—which affects plants indirectly by influencing nutrient availability—can strongly enhance nutrient uptake and plant growth.[32]

Finally, ecologists have been debating since the mid-1990s whether the sheer number of species in an ecosystem plays a fundamental role in how the ecosystem operates. The debate is far from academic. Just as in the Smokies, human pressures everywhere are eliminating distinct populations of plants and animals and threatening to drive vast numbers of species extinct. What else will we lose when they go? Will the ecosystems we rely on for basic life support services falter as species disappear?

The vast majority of experiments testing these questions have focused on a single taxa and a single critical process: that is, the impact of plant species richness on productivity. Productivity means the total mass of greenery, roots, and other plant tissue produced on a site; it's the process that fuels food webs above and below the ground, including those that supply us with food, timber, and other essential goods. The results so far demonstrate fairly clearly that more plant species are usually better for maintaining lush growth on a site through good times and bad, but the reasons are hotly contested. It is generally agreed that the identity and talents of the plant species in a community strongly affect productivity levels. Ecologists do not agree, however, on whether the actual number of plant species matters much—except that the more species there are, the more likely it is that the community will contain a few dominant and highly productive ones.[33]

In the soil itself, researchers are increasing their efforts to test the relationship between the diversity of soil creatures—or soil functional groups—and ecosystem functioning.[34] Just as in the aboveground studies, results so far indicate that the talents of individual species and the array of different functional types in a soil community have more influence over ecological processes such as decomposition than does the number of species actually present.[35] Vast numbers of species in the soil perform most jobs, a phenomenon known as "functional redundancy." To get an idea of how many different types of soil animals throughout the world contribute to the decomposition process, for instance, an international team of scientists led by Diana Wall filled 1,000 mesh bags with alfalfa grass and put them out on the ground in 18 countries, from the Namibian desert, Polish fields, and Tasmanian forests to the grasslands of the midwestern United States. After a few months, the participants in this Global Litter Invertebrate Decomposition Experiment collected all the creatures that had crawled up from the soil to shred, tear, grind, or otherwise begin breaking down the litter in the bags and sent them to Australia for identification. By late 2004, the GLIDE team had collected nearly 62,000 soil animals from 37 different taxonomic orders, including earthworms, ants, termites, millipedes, beetles, spiders, mites, snails, thrips, and woodlice.[36]

Different species may perform the same job in slightly different ways, of course, and some functional groups contain fewer players than others, which may make the work they perform more vulnerable to a loss of species.[37] These relatively species-sparse groups include shredders of organic matter such as mites, millipedes, earthworms, and termites; soil-movers such as ants, termites, and earthworms; mycorrhizal fungi; specialized players in the nitrogen cycle such as nitrifying bacteria and denitrifying bacteria; and bacteria that specialize in processing methane, hydrogen, iron, and sulfur.[38]

Even in functional groups with large numbers of species, there is no reason to assume that each can serve as an exact substitute for another or that one can compensate fully for the loss of another. As one soil researcher noted, too often "redundancy may be more apparent than real."[39] We see redundancy because we lump creatures into broadly defined functional groups of our own creation, unaware of the special conditions, hidden talents, strategies, or subtleties in the way each goes about a task. Species we lump into the same functional category, for example, often respond quite differently to disturbances, above- or belowground. Thus, creatures that seem redundant in today's context may prove to be vital backup players as conditions change. Think of it as insurance. This individuality in talents and tolerances also means that we need to go beyond functional groups and learn more about the species underground if we are to understand how soil systems will respond to the kinds of human pressures that threaten the Great Smoky Mountains National Park and too many other places we value: air pollution, acid rain, climate change, the introduction of exotic species, and changing human land use patterns.[40]

As both soil degradation and threats to biodiversity accelerate, learning "who's there and who matters most" will be vital to protecting not only the health of special places like the Smokies but also the integrity of our working lands and waters.

IV

The Power of
Ecosystem Engineers

fter 5 years exploring the world aboard the H.M.S. *Beagle*, English naturalist Charles Darwin retreated to a country home in Kent to ponder all that he had observed and to develop what would become his theory of evolution. Oddly enough, he also began then a lifetime study of earthworms. In 1837, only a year after stepping off the *Beagle*, Darwin appeared before the Geological Society of London and asserted "that all the vegetable mould over the whole country has passed many times through, and will again pass many times through, the intestinal canals of worms."[1] By vegetable mould, Darwin meant what we call humus, the dark, rotting organic matter that harbors much of the nutrient wealth of soil.

Critics quickly objected that worms are far too small and weak to produce such a large impact on soil. This struck a nerve with Darwin. "Here we have an instance of that inability to sum up the effects of a continually recurrent cause, which has often retarded the progress of science," he wrote many decades later, after facing similar objections to his theory that natural selection, acting on slight inherited differences, among individuals over untold generations, gives rise to new forms of life.[2] Darwin would spend more than four decades studying worms, confirming the power wielded by large numbers of little things working over time. Although he did not label the work of worms good or bad, his writings helped burnish the positive image that earthworms enjoy today. It wasn't always so.

Earthworms have endured great swings in reputation through the ages. Aristotle called them the "earth's entrails" because they digested soil and debris, and like the early Egyptians, he believed they promoted soil fertility.[3] But over the centuries, it seems, worms slipped in farmers' estimations into the ranks of pests and destroyers of crops. By the 18th century, English parson and naturalist Gilbert White felt the need to defend earthworms against the "detestation" of gardeners and farmers—gardeners because they lamented the "unsightly" mess worm castings made of their garden paths, and farmers because they believed wrongly that "worms eat their green corn." In truth, White wrote, "the earth without worms would soon become cold, hard-bound, and void of fermentation; and consequently sterile."[4] Today, earthworms receive little but good press in popular gardening magazines. They are among the few soil creatures that almost everyone recognizes and almost universally regards as "good"— industrious and benign icons of soil fertility. A sign I noticed on one museum exhibit on biodiversity summed up the popular wisdom: "If you spy a lot of earthworms in the ground, you're probably looking at healthy soil."[5]

That's why, when Cindy Hale gets up to talk about earthworm damage in Minnesota's maple forests, her audiences usually respond with open-mouthed disbelief. Most of her listeners are as unaware as I was that the worms in our northern gardens, fields, and forests are not natives. There are no native worms across most of the upper swath of North America, I soon learned, a void that some scientists blame on the advance and retreat of Pleistocene glaciers (1.8 million– 11,000 years ago). Our night crawlers, leaf worms, red wigglers, and other common backyard worms are mostly European invaders that hitchhiked here a century or more ago in ship ballast or on the root balls of imported plants and then settled into human-dominated landscapes. In recent decades, earthworm introductions to natural areas have accelerated as the import of bait worms for sport fishing and compost worms for vermiculture (raising earthworms to compost organic waste) has grown into a multimillion-dollar industry. Hale, a research associate at the Natural Resources Research Institute, University of Minnesota, Duluth, for example, has been watching invasion fronts of exotic bait worms fan out from fishing resorts and boat landings into Minnesota's previously worm-free north woods. She explains to audiences how the advancing worms devour the thick duff of the forest floor and make life difficult or impossible for many native wildflowers, tree seedlings, and small creatures above and below the ground.

"After they get beyond disbelief, people wrinkle up their foreheads and ask: 'So, does that mean earthworms are bad?'" Hale recounts. "I say, 'They're not bad or good. Worms just do what they do, and in some places we like what they do and in some places we don't like what they do. It's not a yes or no question.'"

In fact, the more I learned about what earthworms do, the more naïve it began to seem to caricature these complex and influential creatures as helpful or harmful, as I had once done. Although misplaced worms are causing unwanted disturbances in many natural habitats, in other parts of the world researchers are learning to harness their powers to restore fertility and enhance crop production on degraded lands.

Biologists have long described the earth's 3,000–4,000 earthworm species as "bioturbators" because of their earth-churning ways. But worms do more than plow through the soil, swallow dirt, and excrete it some distance away. Worms swallow and break up leaf litter, organic debris, and microorganisms, living and dead. Inside the worm, this fragmented organic material is mixed with soil, digestive juices, and mucus, creating a feast for gut-dwelling microbes, as well as other decomposers that survive the passage. Beneficial microbes such as nitrogen-fixing bacteria and mycorrhizal fungi get moved from place to place by passing through worms, too, as do disease agents and their natural enemies. Worms cast or excrete enriched, gummy soil in the form of tiny clumps or aggregates, creating hotspots

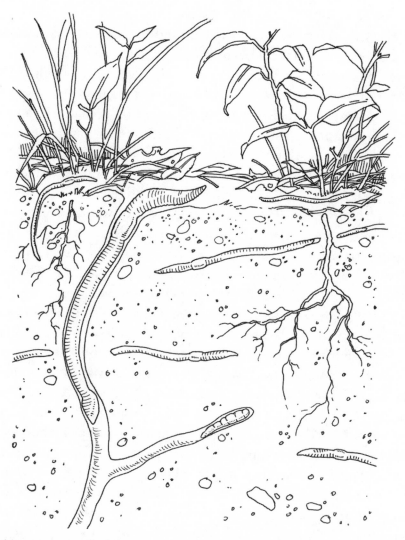

Earthworms are called "ecosystem engineers" because their burrowing not only churns the soil but also changes the habitat of other plants and animals, for better or worse.

of biological activity and enhanced nutrient cycling that influence the composition and perhaps the species diversity of the soil community for weeks or months. Further, as worms burrow and feed, they usually increase the porosity of the soil, aerating it, improving its structure and water-holding capacity, and creating channels for water to drain and roots to penetrate. Thus, earthworms are not just biological plows. They create and modify habitat and alter the resources available to other creatures in their sphere of influence—called the "drilosphere"—just as plants do in the rhizosphere, the neighborhood adjacent to their roots where microbes thrive on exuded sugars.[6]

Although ecologists have long focused on predation, competition, parasitism, and symbiosis as key interactions that shape ecosystems, habitat creation was not singled out for study until the early 1990s. That's when ecologists came up with a descriptor for creatures that engage in it: "ecosystem engineers." With that, earthworms took their place amid the ranks of beavers, elephants, woodpeckers, prairie dogs, gophers, ants, termites, burrowing shrimp, and other earth-moving creatures whose routine activities both transform the character of the land and the seabed and shape the vital ecological processes that take place above and below ground.[7]

Although Darwin didn't use the word, it was the engineering work of worms that fascinated him most. He purposely placed chunks of stone on the surface of a pasture near his house and waited 29 years to see how deeply worms would bury the material by bringing up deeper soil layers and casting it on the surface (7 inches, he learned, or one-quarter inch per year). Similarly, with a remarkable detachment few of us could muster when it comes to our own landscaping, Darwin watched for more than three decades as a flagstone path in his lawn sank and finally disappeared under layers of worm castings. Meanwhile, he meticulously monitored the actual weight of castings brought up each day from individual worm burrows. In the end, Darwin concluded that worms working underground were swallowing and bringing to the surface more than 10 tons of earth every year on each acre of English countryside.[8] More than a century later, Patrick Lavelle of the Laboratoire d'Ecologie des Sols Tropicaux in Paris found that earthworms in the tropics can ingest as much as 200–400 tons of soil per acre per year.[9]

Lavelle and others recognize that soil engineers—and indeed, all soil animals—are resources that need to be tended, whether they are used as "therapy" for degraded lands, enhancers of crop production, or indicators of the ecological impact of agriculture and other human land uses. Like human engineering activities, however, the dams, tunnels, and earthworks of soil engineers—especially misplaced ones—can sometimes degrade rather than enhance our land and water. Hale is one of a growing number of ecologists, foresters, and land managers who are coming to recognize that releasing powerful creatures in the wrong places—or eliminating them from their native habitats—can wreak unwelcome changes in ecosystems we value.

Hale's answer to the uninitiated—school children, foresters, ecologists, and skeptical writers alike—is to invite them into the sugar maple woodlands of Chippewa National Forest, a popular fishing and boating area in northern Minnesota. That's where I meet her one mid-morning in late June, in the little town of Cass Lake that serves as the national forest headquarters. At that time, she is just finishing up her doctorate on the impact of worms on these hardwood forests. A Duluth native, Hale is a bit older than most graduate students, the result of what she genially calls her "nonlinear career." After earning an undergraduate degree in ecology, she "went off and did jewelry design for 5 or 6 years, and seasonal temp jobs all over on wolves, owls, plants, anything I could get." When she decided to go back to graduate school, it was to work on old-growth forests in Minnesota. Worms were not on her radar screen then.

Before heading into the woods, Hale and I and another curious writer stop for lunch at a modest resort with rental cabins and a marina. It's Friday, and trucks pulling trailered boats are already streaming past on Highway 2 from Duluth. The bait shops around Minnesota's 14,000 fishing lakes will do a brisk business this summer weekend in night crawlers and leaf worms (also called red worms), two European species known formally as *Lumbricus terrestris* and *Lumbricus rubellus*. By Sunday afternoon, thousands of fishermen will be trailering their boats again and dumping any leftover worms onto the ground near marinas and boat launch ramps. Unless, that is, they've seen the "Contain those crawlers!" posters Hale has been distributing. Her message is simple: Dumping worms in the woods harms the forest, and it's illegal. Dispose of leftover bait in the trash.[10]

It's the seemingly harmless act of bait dumping, repeated every weekend for decades, that has set in motion an unwelcome change in these woods, Hale recounts over lunch. Starting in the 1980s, soil scientist Dave Shadis and a handful of other biologists working in the Chippewa National Forest had noticed changes in the understory plant community at several sites. Maple seedlings were growing sparser, and so were ferns and wildflowers such as large-flowered trillium, yellow violets, bellworts, Solomon's seal, and wild ginger. It wasn't until 1995, however, that Shadis thought about worms as culprits. That year he read an article about the disappearing understory in New York's urban woods.[11] The changes were occurring in wooded parklands overrun by earthworms. When Shadis went out to his own affected forest sites and dug, he found worms. In places where the understory remained unchanged, he found no worms.

A few years later, Shadis led a group of ecologists on a field trip to see the contrast between worm-free and worm-infested woods. Hale, then a master's student, went along.

"Everybody was just blown away," Hale recalls. "Because most of us didn't even know worms weren't native or that they could have negative impacts. That's completely contrary to everything we've ever learned since we were this high"—she stoops to hold her hand at knee level—"about worms being these benevolent little creatures who just spend their time serving us."

When it came time for her to pick a Ph.D. topic in 1998, Hale remembered Shadis's observations and decided to pursue the worm connection with the help of her advisor, Lee Frelich, director of the University of Minnesota Center for Hardwood Ecology.

"It's been rewarding," says Hale, who plunged into public education about worms as eagerly as she embraced the research. "The first few years, we were like voices in the wilderness. It's been as much a sociological experiment as a biological one, seeing how people—not just the lay community but the scientific community—react to this research and the implications of it."

After lunch, we drive east on Highway 2, then turn south down a bootleg-shaped neck of land called the Ottertail Peninsula that juts from the north shore into Leech Lake. We park at the edge of a wood and Hale begins unloading her gear: a long-handled garden bulb planter she uses for taking soil cores, a foot-square metal frame, two plastic jugs, and a large tub of ground yellow mustard purchased from a food co-op in Duluth.

The western shore of this peninsula is lined with fishing resorts built between the 1920s and the 1950s that are the source, apparently, from which the worms have been fanning out into the woods, she says.

"That's what we see across this region, a very distinctive pattern of these leading edges of worm invasion radiating out from sites like boat landings, fishing resorts, things like that," she explains as she fills the jugs with a dilute mustard and water solution. "People who had been watching the change in the understory vegetation—although they didn't link it to earthworms at the time—noted that the change jumped this road about 15 years ago. And now it's about 300 yards into this stand. So we're going to walk in, cross what we call the visible leading edge—which is where you start to see forest floor again—and then beyond to what we consider relatively worm-free conditions with thick forest floor and lush native plants."

By "forest floor," Hale means what is variously called vegetable mould (in Darwin's day), leaf mold, compost, duff, or among soil scientists, *mor humus*. It's a spongy, springy surface layer composed of many seasons worth of rotting leaves, twigs, bark, animal remains, and other detritus. If you push aside the identifiable leaves and other crispy bits in the upper inch or so, you will find a dark, slippery mat of skeletonized leaves and other older material that is slowly losing its identity and substance to the feeding of a threadlike network of white fungal hyphae. Worm-free sugar maple forests are known for their slow, thrifty nutrient cycling, and the carbon and nitrogen in a fallen leaf may sit locked away in the duff for 3–5 years like banked wealth before soil animals and decomposer microbes break the material down and release the nutrients for reuse.

Since nothing was known about the actual worm populations here when Hale began, her strategy was to pick several sites in Chippewa where the forest floor and the plant community were visibly changing, then mark off a series of plots at each site along a 500-foot transect running from the changed into unchanged forest. Ever since, she has been documenting shifts in the worm populations, forest floor, upper soil horizons, and understory plants as the worms march on. In the first 4 years of her study, the invasion front of worms advanced as much as 100 feet.[12]

It has rained heavily for the past week, and now the midday sun is enveloping us in steamy heat. We gladly follow her into the cool, dappled shade of the sugar maples. A thin layer of soggy leaves that dropped last fall still coats the bare ground. Hale predicts most of this litter will be gone in a few weeks. The worms at this heavily invaded end of the transect rev up the pace of nutrient cycling, consuming the whole year's leaf fall and incorporating the carbon and nutrients into the soil during the 2–3 months they are active. (Worms in Minnesota hunker down in the soil or litter and wait out the frigid winters. In the droughty summer months, some go into a torpid state called estivation—similar to hibernation, but in warm weather.)

A little farther into the woods, Hale kneels down and brushes aside the leaves.

"Underneath you see tons and tons of earthworm cast material —or worm poop—these granular, kind of globular piles of soil. You can see a lot of worm burrows exposed to the surface." She scoops up some material and holds out a 1-inch glob. "This is a piece of a midden, literally just a pile of cast material that night crawlers form around their burrow. And it often has lots of these little tufts of leaf petioles and veins because that's the remnants of the leaves that they've ingested and pulled down into their burrow."

She stands and pushes the bulb planter into the cleared spot with her foot, pulling out a 7-inch core of soil. She removes it from the metal tube and holds it up for us to see. "The A horizon is this black upper layer, about 5 inches thick," Hale explains. "Then below you can see it starts to grade slowly down to this buff-colored soil. This is the top of the E horizon where the organic materials are leaching down from this A horizon." At the worm-free end of this transect, she says, we will see a thick O (organic) horizon—the forest floor layer—

then a very thin or no A horizon atop a very thick, light-colored E layer. When worms arrive, they consume the organic layer, churn and work it in their guts, mix it with mineral soil, and excrete what becomes a thick A horizon.

A few days later, in an old-growth maple and basswood (linden) forest preserve called Wood-Rill just west of Minneapolis, Lee Frelich would tell me that soil scientists long mistook the thick A horizon at that worm-infested site for a "plow layer."

"They think these woods have been logged and farmed and abandoned back to woods," he recounted. "But as forest ecologists, we knew this site hadn't been logged. So how do you get a plow layer without plowing? Well, the answer is, earthworms are plowing." Frelich has spent most of his career studying the effects of fires and windstorms on forests. His attention turned to worms when the landowner who donated Wood-Rill as a preserve asked what had happened to the wildflowers he enjoyed as a child. "So I came out here and looked and discovered it was the worms," he recalled. "Then I had to get into the worm thing because obviously, we couldn't ignore it."

Both Frelich and Hale believe the worm invasion went unnoticed by land managers for too long because people trained in forestry seldom recognize changes in soil organisms, while soil scientists often don't know what the plant community should look like.

The dark soil Hale is holding out now looks to me like what I'd want in my garden, and I say so.

"That's exactly it," she agrees. "In fact, when I give talks to people, they say 'that looks like good, rich black garden soil.' And I say 'you're right,' because that's exactly what it is. But that's not what's supposed to be here in the native worm-free condition. And it changes everything, changes all the ground rules of the ecosystem. A lot of people say, 'but this is really good soil, why don't the native plants do better in this soil?'"

The general answer is that native plants have developed in the absence of earthworms at least since the retreat of the glaciers.

"So most of the native understory plants here root almost exclusively in the forest floor layer [organic horizon], and many have very complex seed germination and dormancy strategies. Some may take two or three freeze-thaw cycles for full germination, and during that time the seeds have to be protected from desiccation, freezing, predation. A forest floor does that really well."

As the total mass of earthworms has increased on her sites over the years, Hale has watched the abundance and diversity of wildflowers and other herbaceous plants as well as the density of tree seedlings plummet. Sites where the forest floor was once at least three-fourths covered with lush greenery are now three-fourths bare ground. Earthworms have been primarily responsible, but an over-abundance of deer has also taken a toll. Hale and Frelich theorize that some of the more robust native plant populations knocked back by the arrival of earthworms might eventually rebound, even in the face of heavy deer grazing. Some plants, such as sedges, even thrive with worms because, unlike most native plants, they do not depend on a partnership with mycorrhizal fungi, which can be disrupted by the activities of earthworms. Rare plants, on the other hand, may drop to such low numbers that they are driven locally extinct by the combination of worms and deer. The rare goblin fern, for example, a species that grows mostly between the duff layer and the mineral soil and does rely on mycorrhizae for sustenance, is likely to be completely eliminated from worm-invaded sites.[13]

We stop to look at some stalks rising from the ground. "These are wild leeks, one of the first things to come up in the spring." Hale pushes aside the sparse litter to reveal a small white bulb poking partially above the soil. The half-exposed leek bulb reminds her of another anomalous reality here.

"Contrary to everything you've heard about earthworms, we actually see an increase in soil compaction and bulk density [the weight of soil in a certain cube of space] as a result of earthworm activity," Hale says. "There's almost a doubling in bulk density in these sites relative to the worm-free sites."

I look skeptical. What about all the burrowing and casting that's supposed to increase the pore space and fluff the soil?

"These native woodland soils are so light and so low density that the earthworm cast material is much more dense," she responds. "So we actually get an increase in compaction. As the forest floor is eaten, the soil sinks anywhere from 4 to 6 inches and you get exposure of those root crowns. We see a big increase in sapling mortality because of it as earthworms invade."

Sure enough, as I turn around I see a large yellow birch, its roots snaking atop the soil surface like the roots of tropical forest trees. At Frelich's Wood-Rill site, which lost its protective forest floor decades ago, the bare worm-worked soil is plagued by erosion as well as compaction.

The effect is not confined to temperate forests. In central Amazonia, where tropical rain forests have been cleared to create cattle pastures, the soils quickly lose two-thirds of their original "macrofauna" species—including ants, termites, millipedes, spiders, mites, beetles, and native earthworms—and are overrun by dense populations of an aggressive nonnative earthworm, *Pontoscolex corethrurus*. The pasture soil, already compacted by the heavy machinery used in clearing the forest and by the trampling of cattle, gets dramatically denser as it is passed through the guts of worms at the rate of about 40 tons per acre each year. Patrick Lavelle and his colleagues found that without the "decompacting" activities of ants, beetles, and other soil animals to break up the cast material into smaller granules and restore porosity, the soil becomes impervious to infiltration by air and water and discourages plant growth.[14]

Hale picks a spot and sits down, brushing aside leaves again to reveal bare soil. She pushes the metal frame she carries into the soil an inch or so and slowly pours in about a half gallon of the milky-yellow mustard water she's carried with her. Within seconds, the ground is squirming with small, irritated worms. Most of them are juveniles and hatchlings only one-quarter- to one-half-inch long. They hardly seem the sort to drive large ecological transformations. She begins picking them up with tweezers and dropping them into a shallow plastic dish of alcohol.

Earthworms are difficult to identify to species until they're sexually mature, so Hale usually samples in fall when a larger proportion of these worms will have reached adulthood. After picking out two dozen from the first flush of worms, she pours on a second dose of mustard water. The rain has brought a fresh hatch of mosquitoes that dart at our necks and faces as we sit staring at the new crop of worms surfacing in the extraction frame.

Hale and her colleagues have found seven species of earthworms invading here, all members of the European Lumbricid family. Together they cover the three basic ecological groups of earthworms. First are the litter-dwelling or epigeic species, worms you'll often find in your compost heap but seldom in your garden soil. They're small-bodied red-brown worms that live in and feed upon the litter layer or near the surface of the mineral soil. At this site, the epigeics are *Dendrobaena octaedra*, *Dendrodrilus rubidus*, and *L. rubellus*, the aptly named leaf worm.

Second is a single species of anecic or deep-burrowing worm, the night crawler *L. terrestris*, now found throughout most of the world. (The traditional term for widely introduced worms such as *L. terrestris* and the tropical *P. corethrurus* is "peregrine" species—essentially, wanderers.) Anecic species are usually very big, pigmented worms that live in permanent burrows as much as 6 feet deep and feed on fresh surface litter that they pull into the burrow opening and mix deep into the soil profile.

Third are the endogeics, meaning "in soil," which form lateral-branching tunnels 15–18 inches below the surface as they feed. The endogeics here are larger bodied but nonpigmented species in the genera *Aporrectodea* and *Octolasion*. (Tropical *P. corethrurus* is also endogeic.) Endogeics consume soil and feed directly on the organic matter it contains, as opposed to fresh litter, so you seldom spot them on the surface. Yet gardeners who pull a plant up by the roots will often see *Aporrectodea caliginosa*, the common grayish pink field worm, coiled in the roots.

Hale fishes with the tweezers in the plastic tray, trying to show us how to tell the worms apart. I find it difficult to distinguish even the colorless endogeics, however, because the dark soil in their innards shows right through their transparent "skin" or cuticle. These beasts possess no eyes or ears, yet their segmented bodies are studded with light receptors and they flinch at touch or vibration.

"This is a sexually mature adult worm. It's got the clitellum, that smooth little band or necklace that you think of when you think of earthworms." She holds up a 2-inch worm and points out the smooth saddlelike patch near the head end. For those who know what to look for, the clitellum can be used to tell worm species apart. It's also important in worm sex.

All Lumbricids are hermaphrodites, meaning they have both male and female equipment. Some can self-fertilize but many mate sexually, lining up head to toe and encasing themselves in a slime tube secreted by the clitellum to help maintain what can be an hour-long embrace. With each worm playing both male and female roles, each later produces a cocoon using a layer of skin sloughed from the clitellum.

Hale gathers up her equipment and we move on toward the less invaded end of her tract. She once brought a group of resource professionals out here and showed them what we've just been seeing: the absence of a forest floor, the scarcity of maple seedlings, and the abundant sedges. "A couple of them said, 'but this is what most of the forests look like,'" she recounts. "Many of these were people from southern Minnesota, which is much more heavily impacted by human activity and has very few worm-free sites remaining. And when we finally got up ahead to the worm-free site with the thick forest floor, they just shook their heads and said, 'this makes me realize, I may never have seen a worm-free site.' It was an incredibly powerful revelation to realize that all of their impressions of what a natural forest looks like may be based on something that is heavily impacted by an exotic species."

As we follow the worms forward, we begin to spot a scattering of trillium, blue cohosh, spikenard, and yellow violets, flowers that have virtually disappeared behind us. Here, Hale points out, "there are lots of little two-leaf maple seedlings. In many of the worm-free areas we can get sugar maple seedling densities of 100–200 per square meter [slightly larger than a square yard], as opposed to one or two in the worm-impacted areas."

Hale takes another soil core from this recently invaded spot. It shows a thin remnant of forest floor, little more than an inch of A horizon, then a long plug of fine silty beige soil. The thinning of the forest floor doesn't bode well for the future of the wildflowers around us. Meticulously, she pokes the divot back in its hole and moves ahead.

The ground becomes spongier as we near the end of her transect, and Hale looks for a clearing amid the plants and seedlings to set up her frame and extract more worms. Here the duff layer, the organic horizon, remains several inches thick. She pushes it aside, clearing a spot for the metal frame, and pours on a whole gallon of mustard water. Within seconds, tiny white *Dendrobaena* are thrashing on the surface. Even the adults among them are less than an inch long.

"Well, even though I euphemistically refer to this as the worm-free end, it's really not worm free," Hale sighs. "But there's a smaller suite of species, biomass is lower, and it's been invaded for a much shorter period of time."

It's also not surprising to Hale that the worms we're seeing here are litter-loving *Dendrobaena*. It turns out that worm species invade in predictable succession in this forest, and tiny *Dendrobaena* takes the lead. Hale has also learned, both from this field study and from greenhouse experiments, that different worm species create dramatically different ecological impacts. In these sugar maple forests, the arrival of *Dendrobaena* has almost no impact on forest floor thickness or on understory plants. "But when the leaf worm *L. rubellus* shows up, we see very rapid removal of the forest floor and also bigger declines in native plant populations. These worms literally eat the forest floor out from underneath the roots of the plants." The species declines to low levels as it destroys its own habitat, leaving the later arriving night crawlers and the endogeic species to dominate.

Despite the accelerated pace of decomposition and nutrient turnover spurred by these invaders, Hale's studies turned up another counterintuitive result: the amount of nitrogen available to fertilize plant growth actually declines in the worm-worked soil here. Her studies don't explain why, but others have found that earthworm activity can increase nitrogen losses, either in the form of nitrate leaching into groundwater or as gaseous nitrogen and nitrous oxide released to the air by microbes.[15] Even in crop fields where farmers value many earthworm species for their positive effects on plant growth, there is a suspicion that deep-burrowing anecic worms such as night crawlers can increase the movement of nitrates, pesticides, and other pollutants down through the soil and into groundwater.[16] Indeed, a recent study in an Ohio cornfield found that nitrogen leaching from plots with high earthworm populations was 2 1/2 times as great as in plots with low levels of worms.[17]

In places where invading earthworms don't cause detectable leaching or even speed up nitrogen cycling, the worms still can dramatically alter the way the system handles soil carbon, nitrogen, and phosphorus. The clear message is that ecosystems with earthworms—especially those where worms were previously absent—can work quite differently than ecosystems without worms.[18] How a forest or other ecosystem responds to earthworm invasion will vary with the species of worms involved, the nature of the ecosystem itself, environmental conditions at the site, and even past land uses. Finally, the activities of worms will interact in complex ways with other human pressures—acid rain, climate change, exotic pests and diseases, timber and farming practices—that are also driving changes in our lands and waters.

Exotic worm impacts are increasingly drawing the attention of researchers and land managers, from the maple forests of upstate New York to the aspen forests of Canada, the oak savannas of California, and the boreal (subarctic) spruce-fir forests of Russia. Hale and Frelich have expanded their inquiries into boreal spruce and aspen as well as beech hardwood forests. Meanwhile, Paul Hendrix of the University of Georgia is leading an effort to see what happens in regions where European and Asian earthworms encounter some of the 90 or so species of native earthworms in the United States, from North Carolina to the Oregon coast. So far, the most dramatic effects of exotic worms, at least in North America, have been seen in areas long devoid of native earthworms.[19] Yet newer arrivals with different traits have the potential to transform both native earthworm communities and areas long ago reshaped by European earthworms. A whole suite of Asian worms in the genus *Amynthas*, for instance, is now spreading into natural areas, from New York to the hills of Georgia and even the borders of Great Smoky Mountains National Park.

"This is the next problem in the northeast forests," Patrick Bohlen told me, referring to *Amynthas*. A researcher now at Archbold Biological Station in Lake Placid, Florida, Bohlen previously investigated the impacts of European worms on soil processes in New York maple forests.[20] Now Asian worms are spreading toward these same forests from urban areas where they arrived either in the soil of potted plants or as contaminants in shipments of bait or compost worms. Unlike European worms, which usually slip into new territory without drawing the immediate attention of the general public or even scientists, *Amynthas* species make a dramatic entrance.[21]

"Their casts are very distinctive, and they really transform the forest soil surface into a pile of crumbs," Bohlen says. "Their behavior is distinctive, too. If you pick them up or disturb them, they move like wiggly snakes or flip in an S shape." *Amynthas* are annual species that hatch in spring and die in fall, so to perpetuate themselves they reproduce in huge numbers. The soil can literally appear to be writhing with worms. "When people encounter it, they very often react to these worms as to pests," Bohlen says. "They're aghast. It's off-putting."

What effect *Amynthas* will have in the long run is anyone's guess. Ecologists and foresters are still trying to follow the ripple effects of European worms through invaded communities. Some of the long-term effects of invasion are visible in places such as Wood-Rill—bare, eroding soil, sparse understory, and minimal seedling regeneration under centuries-old trees. And there are the unseen effects such as reduced nitrogen availability. No one knows yet what this means for the growth rate of trees and the future productivity of the forests themselves, data Frelich considers critical for the region's forest industry as well as for conservation.

As for other soil denizens, worm activities and the loss of the forest floor seem to be particularly hard on fungi, including mycorrhizae, as well as on litter-dwellers such as springtails, mites, and millipedes. Above the ground, species such as ovenbirds that nest in thick forest floors are expected to suffer, along with birds that require a lush understory. Already, loss of the forest floor in Chippewa has reduced the numbers of small masked shrews and red-backed voles that forage for insects, seeds, and fungi in the duff; yet deer mice and wood mice appear unaffected.[22] In the worm-invaded woodlands of New York, adult red-backed salamanders seem to benefit by feasting on exotic earthworms.[23]

Many ecologists worry that the real beneficiaries of exotic worms, however, will be other exotic creatures that further alter the soil as well as life aboveground. In Hawaii, for instance, exotic worms have spread into the native forests, providing a protein source for feral pigs that physically damage native vegetation and spread the seeds of exotic plants. It's a synergistic onslaught that leads to what one ecologist has called "invasional meltdown."[24] In the forests of New Jersey, researchers find higher nitrate concentrations as well as higher densities of European earthworms under invasive barberry shrubs and wiregrass than in the soil under native shrubs.[25] The same holds true for invasive buckthorn shrubs in the Chicago area.[26] Conservation managers who try to restore invaded areas like these by removing barberry or buckthorn and replanting native shrubs may find their efforts foiled from below by the legacy of changes in soil nutrient cycling and the worm-dominated soil community.

"The more you think about it, the cascades of potential effects are really dramatic," Hale comments. What can be done to halt the worm invasion or restore the invaded ecosystems? I ask. Earthworms cannot be eradicated once they invade, but Hale believes we can protect uninvaded sites and slow the advance of worms elsewhere.

"Expansions of established earthworm populations are really quite slow, maybe 30 feet a year," she points out. "If you do the math, it takes a couple hundred years to go a kilometer [two-thirds of a mile]. So if we can prevent new introductions in sites that are still worm free, we can buy ourselves hundreds of years to find solutions. The education component is very important here."

Another useful management strategy is to prevent new species from joining the mix of invaders. "People say, 'if there are already earthworms invading a site, why does it matter if I release earthworms?' And my answer is that we might not have the whole suite of species, and the type of impact you get depends on the species assemblage." Minnesotans should even be wary of Asian *Amynthas* species, which apparently don't survive the region's harsh winters, Hale says. A warming climate could remove that limit.

Finally, reducing other pressures on worm-invaded ecosystems might help native plants and animals survive.[27] One key pressure in Minnesota is deer grazing, as alluded to earlier. "We're not convinced that many of these native understory plant species are incapable of co-existing with earthworms," Hale says. But the intensity of deer grazing means plants knocked back by worms get little opportunity to recover. Frelich's team fenced off two 1-acre areas at Wood-Rill to exclude deer and within a few years saw blue cohosh, bloodroot, trillium, bellworts, and other plants nearly eliminated outside the fence by the worm invasion springing up from seeds long dormant in the soil. "These plants have probably been germinating for decades," Hale believes, "but as soon as one little green thing pops up from the bare ground, the deer eat it."

Inside the deer-free enclosures, Frelich's team is also testing whether electroshocking the ground to remove most of the worms from small patches will eventually allow the forest floor there to recover and nutrient cycling to return to preworm dynamics. It's an interesting experiment—one that Frelich expects to continue for 20 years or so—but an unlikely management option. I asked Frelich what would be the best he could realistically hope for at Wood-Rill.

"I hope the result of this experiment is that the plants grow quite well in the presence of earthworms but without deer," he offered. "Because if it turns out all you have to do is control the deer, that's a whole lot easier than getting rid of the worms."

I've spent much of this chapter exploring the deleterious effects of earthworms because these accounts have helped to jar my own stereotypes about worms and provide vivid illustrations of the power soil animals can wield in shaping the world we experience. Yet I wouldn't want to tip the reputation of earthworms back toward 18th-century "detestation."

Over the past century, earthworms—mostly Lumbricid species in temperate regions—have gotten generally good marks from agricultural researchers and promoters of organic gardening for their ability to enhance plant growth. Two cases in particular have become textbook classics. One involves Dutch "polders," pastureland reclaimed from the sea by diking and draining in the 1950s and initially devoid of earthworms. Intensive grazing by cattle and sheep quickly compacted the young soil and damaged grass production. Night crawlers and other Lumbricids were introduced to some of the degraded pastures in the early 1970s. As the worms slowly spread, they incorporated the mats of dead grass on the surface into the soil, speeding development of an A horizon, aerating the soil, improving water infiltration, and enhancing grass production.[28] In the second case, European worms introduced into New Zealand pastures also dramatically improved grass yields.[29]

In recent decades, several hundred studies in the tropics involving a wide assortment of earthworm species, crop plants, and soil types have demonstrated that worms usually boost plant growth. A review of these studies by George Brown of Brazil's Embrapa Soybean and a number of colleagues revealed that increases in plant growth average nearly 60 percent, and increases in grain yield for crops such as rice and maize average 36 percent when worms are added. The biggest growth enhancements show up in some tropical tree crops and tea bushes, as well as panic grass, an African grass widely planted for forage in the American tropics. Other plants such as oats gain little, however, and the yield of cowpeas, peanuts, and cassava drops in the presence of worms.[30]

Such findings have led to new work for earthworms. For example, Lavelle and Bikram Senapati of India's Sambalpur University have developed an earthworm treatment for degraded soils that has dramatically boosted both yields and profits on aging tea estates in southern India. Some of these plantations have been operating for a century or more, and despite increasing use of fertilizers, the soils suffer from declining organic matter levels, reduced water-holding capacity, acidification, compaction, erosion, nutrient leaching, and up to 70 percent loss of soil organisms—including most native earthworms. Senapati and Lavelle's "bio-organic fertilization" technique involves adding organic waste—various combinations of tea prunings, cow manure, and compost—in trenches dug between the tea rows, along with inoculations of exotic *P. corethrurus* and a mix of other earthworms. Rejuvenation of the soil by this method has increased tea yields from 80 to 276 percent on various estates, and the technique is now being applied to tree and shrub crops in other countries such as China and Australia as well as India.[31]

Earthworms dominate the world of soil engineers, but they are rare in arid regions. In drylands and a number of other regions, termites, ants, beetles, millipedes, and a diverse suite of other native soil animals head the ranks of soil turners and engineers, often producing a variety of complementary effects such as the compacting and decompacting activities already mentioned. Termites and ants that excavate subterranean galleries and nest chambers and transport litter and plant material underground—and termites that feed directly on soil, like endogeic earthworms do—profoundly influence the structure and the flow of materials and energy in the soil.[32] According to Bert Hölldobler and Edward O. Wilson of Harvard University, "One third of the entire animal biomass of the Amazonia terra firma rain forest is composed of ants and termites, with each hectare [2.5 acres] of soil containing in excess of 8 million ants and 1 million termites."[33]

Unlike earthworms, other soil animals are seldom deliberately introduced to rehabilitate degraded land. Increasingly, though, scientists are experimenting with ways to take practical advantage of native soil biodiversity. In part, that means altering practices such as plowing, pesticide use, depleting organic matter, clearing forests, and heavy grazing that are generally harmful to the soil community, especially larger soil animals. Earthworms, termites, and ants, for instance, are all terribly sensitive—though in different ways—to changes in the intensity of human land use at the margins of tropical forests. A major effort is now under way in fields and pastures carved from tropical forests to improve the sustainability of subsistence agriculture as well as protect biodiversity.[34] In sites where land is already degraded, other researchers are devising ways to harness the rehabilitative powers of soil animals such as termites by creating conditions that lure them back to work.

Farmers in the dry tropics often regard termites and ants that forage on grass and plant litter as pests because these animals attack crops, especially when the land has been stripped of all other vegetation. In the impoverished Sahel region of western Africa, however, termites are being encouraged to perform a service similar to that provided by earthworms on Indian tea plantations. Continuous farming along with overgrazing and trampling by cattle in this region along the southern edge of the Sahara desert has left much of the surface bare and crusted, impervious to water and unable to support plant life. Inexpensive and low-tech methods of soil rehabilitation are urgently needed, and native termites, it turns out, are up to the task. Researchers find that when crusted soil is mulched with woody material, straw, or cattle dung, termites quickly arrive to consume it. These are mostly "higher" termites in the subfamily Macrotermitinae that can carve out miles of subterranean galleries per acre, drawing organic matter into the soil, breaking up the surface crust, increasing porosity and water infiltration, and allowing plant roots to penetrate. Within only a year, native plants reestablish on the denuded land and crops such as cowpeas yield modest harvests.[35]

Such successes are testament to Darwin's insight that large numbers of little things have the power to alter landscapes. Certainly scientists and land managers today cannot afford to discount, as Darwin's contemporaries did, the potential of earthworms and other soil animals to reshape the world we experience, for better or worse.

V

Plowing
the Seabed

The 65-foot research vessel *Squilla* is idling in the lee of a stone jetty that shelters Plymouth Sound from the turbulent waters of the English Channel. On her aft deck, four men are hovering around a massive stainless steel tripod suspended from the *Squilla*'s deck crane, ignoring the rain sheeting off their yellow rubber slickers and coveralls.

"It's a boy's toy," Melanie Austen quips as we watch from the partial shelter of a bulkhead. Austen is a marine ecologist at Plymouth Marine Laboratory, and two of the hooded yellow figures are her graduate students. The *Squilla* has brought us out into the sound this August morning from the Barbican, a gray stone wharf that has served the city of Plymouth on England's southwest coast for more than 300 years. It was from the Barbican that Charles Darwin set off aboard the *Beagle* and Robert F. Scott set off to the Antarctic. Austen and her students have set off to explore another little-known frontier barely a mile from the wharf.

The tripod suspended just above *Squilla*'s deck cradles a box corer about a foot cubed. One of the students, Mike Townsend, and postdoctoral researcher Dave Parry, along with two crewmen who are bantering in a Plymouth patois the researchers call "barbicanese," are preparing to hoist the whole apparatus over the side and drop it to the floor of the sound where the steel box will sink about 18 inches into soft mud. When the team begins to winch the corer upward again, a steel plate will swing into place under the box and deliver us an intact chunk of the seafloor. A chunk of dark ooze teeming with life, Austen assures me.

"What you think of as plain, boring mud has got quite a lot of things living in it," she asserts. She is talking about what ecologists call the "benthos," the plants and animals living on and in the sea bottom. Because oceans cover more than 70 percent of the globe, the submerged sediments we will be coring today constitute a sample of the most extensive ecosystem on earth. They also harbor one of the earth's richest animal communities. Some 100,000 sediment species have been identified, but that may represent less than 1 percent of the creatures living in the sand, gravel, and mud—mostly mud—of the ocean floor.[1] There may be 100 million nematode species alone in the abyssal ooze, and 500,000–10 million species of deep-sea "macrofauna"— medium-sized animals such as polychaete worms, burrowing shrimp, clams, and snails.[2] The estimates vary wildly because scientists have only limited samples from which to extrapolate across 137 million square miles of largely unexplored seafloor. Whatever the numbers, it's clear that the benthos remains even more firmly concealed in the "black box" of the sediments than life in terrestrial dirt. To help remedy this, marine scientists from more than 50 countries began in 2000 a decade-long Census of Marine Life, using an array of new technologies to track and identify creatures from nematodes to plankton to tuna. Just as on land, however, the effort is hampered by limited funding and by a dearth of experts who know how to identify and classify organisms, especially sediment creatures.[3]

As with soil creatures on land, life in submerged sediments is increasingly at risk from a variety of human activities: fishing practices such as bottom trawling and dredging—the equivalent of plowing the seabed—as well as rising water temperatures from climate warming, aquaculture, oil exploration, waste dumping, installation of telecommunications cables, introductions of nonnative species, and eutrophication from excess nitrogen and phosphorus washed off the land. (Eutrophication results when excess nutrients cause algal blooms, which fuel a population explosion among microbial decomposers at the seafloor that leads to reduced oxygen in bottom waters.) Austen and her colleagues believe threats to the benthos will eventually make their mark throughout the entire ocean food web, from plankton to fish and whales. She and her students are out here to investigate what else we lose besides living diversity when we destroy seafloor habitat.

The men grab the legs of the tripod to steady the corer as the crane lifts it and swings it over the starboard rail. One of them releases the tension on the cable and the corer drops out of sight below the green surface of the sound. Minutes later, they hoist it back, streams of mud and seawater mingling with rain as they lower the corer onto the deck. They detach the box of sediment from the tripod and slide it across to where Austen and the rest of her team, student Kirsten Richardson and a field technician, are waiting.

Two of them drain the seawater from the top of the box, and then Austen jams a second corer—a stainless steel cylinder that looks like an oversized cookie cutter—into the cube of mud. Once they lift the corer off and clean away excess muck from the outside of the cylinder, they lift it over an empty 5-gallon bucket and let the round plug of mud slurp intact into it. Back in the lab, across the vast grassy sward of Plymouth Hoe from the Barbican, each white plastic bucket will become a mesocosm, a miniature replica of the seafloor world.

Today one of the team's goals is to collect 14 of these sediment cores. Their second objective is to capture any creatures they happen upon—specifically, seafloor "engineers" such as polychaete worms (marine cousins of earthworms), burrowing shrimps, urchins, mollusks, and other macrofauna that stir and aerate the sediments just as bioturbators do on land. In 1891, a decade after Darwin published his treatise on the engineering powers of earthworms, scientists began to investigate the work of their marine relatives—specifically, burrowing polychaetes known as lobworms. Marine scientists have since come to believe that these and other creatures that burrow in sediment

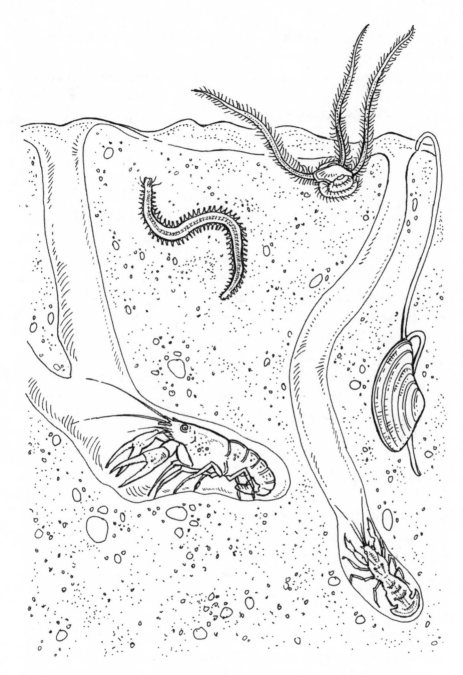

Seafloor creatures such as burrowing shrimps, clams, polychaete worms, and brittle stars stir and aerate the sediments, enhancing the cycling of nutrients that support the ocean food chain.

are vital to nutrient cycling in the oceans, just as bioturbators on land hold significant sway over nutrient processing.

More than 80 percent of all the decomposition and nutrient cycling that takes place on the earth occurs in sediments on the continental shelves and slopes up to 1.2 miles deep, although these areas represent only 16 percent of the area covered by seas.[4] Most of the dead organic matter that is broken down and recycled in these sediments arrives in the form of algae (phytoplankton) that grows on the bottom in shallow areas or that dies and sinks down from sunlit surface waters, along with fecal matter and carcasses of animals ranging in size from zooplankton to worms to whales. Closer to shore, terrestrial runoff including fertilizers and other human wastes can also form an important part of the organic matter input. The nitty-gritty work of recycling all this material is handled by bacteria and fungi, just as in terrestrial soils. These microbes dine on organic matter and release as waste various forms of nitrogen and other nutrients. One microbial waste product, nitrate, is a key fertilizer of the algal growth that forms the base of the ocean food web.[5] One-third to one-half of the nutrients needed to fuel the growth of algae in the seas above the continental shelves is released from the sediments.[6]

Just as on land, however, the activities of larger sediment animals enhance the nutrient cycling process. Like earthworms, marine bioturbators burrow, bulldoze, stir, and "rework" bottom sediments as they feed, increasing the penetration of water, organic particles, oxygen, and other dissolved substances deeper into the sediment, where decomposer microbes do their work. This stirring also speeds the release of microbial wastes such as nitrate from the sediments to the water column.[7] By creating hotspots of microbial activity, bioturbators also attract pinhead-sized animals known as "meiofauna"— nematodes, flagellates, ciliates—that dine on the microbes or their leavings and thus accelerate the recycling of nutrients tied up in microbial cells. Altogether, the presence of bioturbators can as much as triple the metabolism of seafloor communities—that is, the oxygen breathed in and the carbon dioxide released. Only a fraction of this reflects the breathing of the bioturbators themselves; most comes from enhanced activity of decomposer microbes.[8]

What then would it mean for the health of the oceans and the diversity of sea life—including economically valuable fish stocks—if human activities such as bottom fishing greatly reduce or eliminate the work of bioturbators? Are some of these creatures more important than others to sustaining healthy seafloor habitats? How quickly can various benthic communities recover from different types and intensities of fishing? As the United States, the nations of the European Union (EU), and others move toward more comprehensive ecosystem management of fisheries, these questions have become increasingly significant. It is this larger EU interest that drives today's muddy work by Austen and her team.

"Oooh, there he goes," Austen exclaims as her hand darts into the muck of the next core. She pulls out a 2-inch mud shrimp that she spotted trying to make its escape. With her other hand, she pokes around the core surface and pulls out a chunk of the shrimp's burrow, a tube with sides the thickness of a clay flowerpot, slightly brownish in color compared to the dark gray mud from which the shrimp fashioned it. This shrimp is *Upogebia*, an orange-colored suspension feeder that pocks the seafloor with its large burrow openings. Suspension feeders are animals that filter organic particles from the water. Many suspension feeders, including *Upogebia*, some polychaetes, brittle stars, scallops, and clams, nestle into the sediments and poke tentacles, arms, antennae, or siphons into the water to capture food. Others such as sponges, anemones, moss animals (bryozoans), sea squirts, and some crustaceans live atop the sediments.

Austen tells me there is another, smaller shrimp in these waters, the ghost shrimp *Callianassa*—Greek for "beautiful queen"—whose single oversized claw can be nearly half the length of its body. It is a deposit feeder, meaning it actively mines the sediments for organic particles. *Callianassa* also serves the role of a "conveyor belt" for bringing up buried organic matter, foraging deep in the sediment and casting fecal pellets on the surface like a night crawler. As it burrows and feeds, it pimples the seafloor with volcano-shaped mounds.

Shrimp are territorial and will fight if thrown together, so Mike Townsend opens up a Toby Teaboy—an orange plastic tea infuser lined with nylon mesh—and cages Austen's mud shrimp before dropping it into a bucket of seawater. He also drops in a half dozen 1-inch, cone-shaped Turritella snails that he and the others have combed from the mud. Out here where they live and work, these snails are deposit-feeding bioturbators. Most people encounter them in shell shops, however, sliced lengthwise into decorative cross sections for use in crafts and beadwork.

Another box load of mud arrives and Austen points to the poly-chaete burrows and tubes pocking its surface. She fishes in the mud and hands me a 2-inch-long pinkish brown worm called *Nephtys*. Each of its body segments sports a pair of footlike parapodia ending in a cluster of bristles. The worm wriggles sideways in my palm, each segment moving in sequence in an incredibly fluid motion. Indeed, *Nephtys* is commonly called the shimmy worm for this undulating motion by which it swims. Some 10,000 species of polychaetes have been described, but the actual number may be two or three times that. These bioturbators can be found in the top 2–4 inches of almost all marine sediments, usually in great numbers, and are often among the first midsized animals to colonize new or disturbed sediments. Among them are a rich diversity of suspension feeders, deposit feeders, mud swallowers, algal grazers, predators, and even a few parasites.[9]

As the hours pass, the number of shrimp-filled Teaboys and less belligerent creatures in the bucket grows, as does the number of sediment-filled buckets lined up along the stern rail. Everyone is smeared with mud, despite the rain.

"I have friends who think being a marine biologist is a glamorous job," Kirsten Richardson tells us as she tries to rinse the mud from the ends of her plaited hair with cold seawater.

Austen says her 11-year-old nephew should be out here: "He thinks it's a bit exciting having an auntie who's a marine biologist. Of course, it's not always as nice as this when we go out," Austen adds, pushing wet strands of hair from her face. Although she is teasing, she has indeed done this same work in much worse conditions, coring seafloor mud in coastal waters from Greece to Norway, in winter as well as on wet summer days such as this.

Were you one of those students who thought marine biology would be glamorous? I ask. "Oh, yeah! Didn't we all?" she laughs. "Slinging buckets of mud around. I'm sure it wouldn't be many people's cup of tea. But everybody's got their thing. There aren't many people who get to do the glamorous cetacean [whale] work," she says without a hint of envy.

Actually, Austen began her research career focusing on creatures even less charismatic to most people than shrimp and worms: pinhead-sized creatures known collectively as meiofauna, especially nematodes. She still spends part of her time investigating what influences their biodiversity—say, why there are more nematode species in one part of an estuary than another. "I've always looked at ecology in that respect: Why do different communities have different diversity?" she says. "Now this is leading into, does that diversity matter? This is the next step." The question has drawn her into a number of international collaborations and projects, but it has not pulled her away from the mud.

By now, everyone is dishing into the dark ooze as they clean up the cores and hold out offerings to me: brittle stars, a lovely little whelk, a tiny sea cucumber, a half-inch spiny cockle named *Acantho-cardia*, a 1 1/2-inch ottershell clam called *Azorinus* that sports two siphons and makes two holes in the mud, and a larger clam *Lutraria*, which has a single big muscular siphon.

Austen holds up her open palm to show me another polychaete, an inch-long "sea mouse" also known as Aphrodita. I ask why the Greek goddess of love and beauty shares her name with this creature, which looks to me like a hairy little slug. "Because they're really pretty underwater," she says. "Their bristles are iridescent."

Someone offers me another little brittle star, *Amphiura*, a creature that lives just beneath the sediment and waves its arms in the water to capture food particles. Next comes a 1-inch burrowing sea urchin—*Echinocardium*, the heart urchin. *Echinocardium* and a similar Norwegian heart urchin, *Brissopsis*, serve as living bulldozers, pushing their way through sediment as they feed.

Austen rakes her fingers gently through the silky mud atop another core and holds up five more Turritellas. "You can get 10 raking the top like that, so they must be doing something down there," she asserts.

By the time the final core has been slurped into its bucket, the skies are clearing. In the *Squilla*'s galley, we fix tea and snack on ginger cake as the vessel motors a short distance across the sound to Jennycliff Bay where the nearshore bottom is covered with muddy sand. The crew breaks out a small sampling dredge—an iron ring big enough for a person or two to slip through and fitted with a mesh bag formed from several layers of netting. They toss it over the side and the boat chugs slowly in a circle for a few minutes, dragging the gaping mouth of the dredge through the sediment. This dredge is largely for my benefit, although the team also hopes it will yield more creatures to populate their laboratory mesocosms. The crew winches the dredge back up to the rail. The net is full to the ring with mud, so they try hosing it with seawater. As I watch, it seems the mud is being held together by a thick mat of plant roots, as though a large and terribly root-bound potted plant had been tipped out into the net. But there is no garden beneath us. The mat is a tangle of worm tubes made by the polychaete *Melinna*. Dave Parry peels open one of the brown rubbery tubes and pulls out a 1 1/2-inch-long worm.

The vessel moves closer to the shore and the crew drops the dredge again. This time they shovel the catch into plastic laundry baskets, and Austen and her team begin pawing carefully through the muck, intent as treasure hunters at an estate sale. Their efforts yield a tiny sea squirt, a tiny crab, and a literal handful of other prizes. As we motor back toward the Barbican, the sun breaks through the clouds.

About 30 percent of the world's fish catch comes from bottom fishing using towed trawls and dredges as well as immobile or fixed gear such as gill nets, long lines, and various crab and lobster traps.[10] Marine scientists have grown increasingly concerned about the chronic up-heaval created by this industrial fishing gear on seafloor habitats and creatures. Bottom trawls—heavily weighted, bag-shaped nets that can have mouth openings wider than the length of an American football field—are primarily used to catch shrimp and groundfish such as cod, haddock, and flounders. Dredges are rakelike devices with bags—usually with rigid openings—used to collect scallops and clams. Hydraulic clam dredges blast the seabed with jets of water, turning sediments to slurry and floating clams upward to be sieved out by the rakelike prongs or blades of the dredge.[11]

No agency keeps figures on the global extent and frequency of bottom fishing, but one research team estimates that trawlers plow up about 6 million square miles of seabed annually—an area about 20 percent the size of the Atlantic Ocean and 150 times greater than the area deforested by loggers worldwide. In some areas, trawls drag the seafloor multiple times a year.[12] Prime fishing grounds such as the North Sea and the Gulf of Maine, for instance, get trawled at least once a year, disrupting food webs in the sediment and water column over a large part of the world's continental shelves.[13] Increases in the size and power of fishing vessels and mechanization of gear have greatly extended the reach of such trawlers. About 40 percent of trawling now takes place in deeper waters beyond the continental shelves, including slopes, canyons, and isolated submarine peaks known as seamounts, which are extensively trawled for fish such as orange roughy and blue ling.[14]

Historically, the deep sea bottom was considered a nearly lifeless desert, but a landmark study in the late 1960s disproved that notion.[15] Indeed, recent assessments have suggested that the deep sea may support a richer array of species than the continental shelf, and abyssal mud habitats host highly complex communities of polychaetes, mollusks, and small buglike crustaceans such as amphipods (sand hoppers) and isopods (sea centipedes and sea lice).[16]

The direct effects of such fishing on the seafloor have been extensively studied, but the significance of those effects is still hotly debated. Dragging heavily weighted gear over the ocean bottom homogenizes the terrain, reducing natural crevices and hills (but creating new ridges in the tracks of the trawl), crushing or burying worms, sea grasses, sponges, and corals, and eliminating predators such as flatfish, crabs, and shrimp.[17] Further, up to 85 percent of the mass of sea life scooped up in a trawl may be unwanted "bycatch," creatures that are dumped overboard dead or dying.[18] Some marine scientists have compared trawling the seabed to clearcutting forests, a practice that would draw public outcry were it visible to us.[19]

A report by the U.S. National Research Council in 2002—prompted by legislation that now requires fisheries managers to address the impact of fishing on "essential fish habitat" as well as on fish stocks themselves—found that the ecosystem effects of trawling depend, understandably enough, on what type of gear is used in what type of habitat, and how often and how extensively the area is fished. In general, the most vulnerable are stable communities, seldom subjected to natural disturbance and filled with largely sedentary, long-lived species such as corals, sponges, sea grasses, and large clams, as well as gravel and mud habitats. Least vulnerable are communities of mobile, hard-bodied, short-lived species inhabiting naturally changeable environments such as sandy areas swept by bottom currents. The effects of fishing disturbance are cumulative, and the severity depends on the frequency of trawling and dredging. Repeated passes with trawls and dredges may or may not reduce benthic species diversity, but they do drive a change in the types of species in the area, usually a shift from relatively large animals toward small, fast-growing, opportunistic creatures.[20]

Some in the fishing community have argued that this change can be beneficial. Trawling is analogous to plowing a field, the reasoning goes, and in heavily trawled areas of the North Sea the result might even be an increased crop of fish food—mainly worms—along with enhanced stocks of commercially important flatfish such as Dover sole and plaice. But the 2002 National Research Council report concluded that this notion doesn't hold up to scrutiny. Indeed, studies published in that same year found that beam trawling (beam trawls are funnel-shaped nets fitted underneath with many heavy "tickler chains" that rough up the sediments to scare buried flatfish up into the nets) dramatically decreased the number of large animals without affecting the amount of polychaetes that serve as prey for flatfish.[21] What's more, the relationship between mud-dwellers and fish stocks is clearly more intricate than the direct provision of food for bottom-feeders. For example, complex seafloor habitats are known to be important for the survival of many types of fish, providing nursery grounds for juveniles as well as hiding places and food stores for adults.[22] And then there are the invisible ripple effects of such trawling practices on the diversity and productivity of the oceans that are only beginning to draw serious attention.

Back in Austen's office at Plymouth Marine Laboratory, I watch as she pulls up two bright blue sonar images on her computer screen. The left image shows an expanse of seafloor mud pocked with holes—burrow entrances. The right image shows only a series of parallel lines like furrows in a cornfield. The furrows are trawl scars, Austen tells me. No holes are visible.

"Using side-scan sonar, we can see the effects of fishing on the bioturbators visually just like you can on land," she says. It's not quite like walking through a forest and looking for earthworm burrows, but technology is allowing ever greater access to a frontier that is otherwise invisible and inaccessible.

"This is in Norway where we compared trawled areas and non-trawled areas and took samples," she resumes. "It's especially obvious there's a big difference in habitat there. You can actually count the number of burrow entrances before and after trawling from the video. These holes are made by *Calocaris*, a burrowing shrimp that only goes to about 6–10 centimeters [2–4 inches] depth. That's the same depth the trawl will go to in these sediments."

What she is showing me on the screen are the obvious direct effects of the fishing practice in reducing the number of burrow entrances —and presumably the bioturbators that created them. For Austen and her colleagues, the key question is, so what? Do the creatures in the sediment really affect the functioning and the productivity of what's above the sediment?

"That's what we're trying to find out," she explains. "Most people have been asking whether the community of animals on a bit of seabed is the same or different after you haul a trawl over it. People have been accumulating that sort of evidence for a long time. But people haven't really thought about, okay, so what if the benthos changes, is that a bad thing or is that a good thing, and what's it affecting? Why does it matter for the rest of the ecosystem if you have just worms there? We're asking, does that change nutrient cycling? And if it does change nutrient cycling, does that really matter? We want to develop predictions about how primary productivity will change in response to fishing practices in specific fishing lanes from the Aegean to the North Sea. Then from that, we also hope to use fisheries models to see if a change in primary [algal] productivity actually leads to a change in the productivity of the fisheries as well."

Along with research partners in the United Kingdom, Ireland, Norway, the Netherlands, Greece, and Poland, Austen is exploring these questions as part of an EU-sponsored project called COST-IMPACT—"Costing the impact of demersal [near-bottom] fisheries on marine ecosystem processes and biodiversity."[23] The team includes economists as well as marine scientists, and their goal is to use the results of their research and modeling to help managers and policy-makers examine tradeoffs between current fishing efforts and protection of seafloor habitat, biodiversity, ecological processes, and the long-term vitality of the fishery.

It's all part of the drive toward "ecosystem management" of fisheries, Austen says, although that term means different things to different people: "To many of the world's monitoring agencies and government agencies, it used to mean—and sometimes still means—looking at multispecies fisheries instead of just looking at a single fish stock. Or maybe they will look at a single fishery species and its food chain—cod, and what the cod eat, and what eats the cod. For other agencies, 'ecosystem' means looking at social changes, and who's affected by how we manage the seas."

The first step for the research team has been to summarize results from the growing number of fishing impact studies to try to pin down just what *is* likely to happen when a certain type of gear is dragged through a specific type of seafloor habitat, and how long it takes the habitat and the community to recover from that fishing practice. Using a statistical procedure called meta-analysis, the researchers have integrated the sometimes inconsistent findings of 101 different experimental studies conducted in the seas off various parts of Europe, the Americas, Australia, and South Africa and involving different combinations of fishing gear and habitat.[24] It would be another 10 months after my visit before the analysis was complete, but the results broadly confirmed many of the expectations from previous synthesis efforts using fewer studies.[25] The most severe effects on the benthos occur when scallop dredges (metal frames with rakelike teeth and attached chain-mesh bags) are towed through habitats full of corals, sponges, sea grasses, animal burrows and tubes, and other "biogenic" features, for instance, or when dredging takes place in the muddy sand of the intertidal zone (the shoreline area between the high and low tides). On average, intertidal dredging in muddy sand reduces either the abundance of a specific animal species or the total number of benthic species by an astounding 72 percent. It can take years for such communities to recover and for the furrows left by the dredge to be erased. Large, slow-growing sponges and soft corals can take as much as 8 years to rebound after being crushed by a dredge. In contrast, polychaetes in sandy habitats regularly scoured by waves or currents may spring back in a matter of months.

Besides learning which gear causes the greatest impacts in a given habitat, researchers also need to learn more about the work of various creatures that live in harm's way. What happens when you eliminate this type of urchin or that shrimp or worm from a seabed community for months or years? This is where the buckets of mud and animals Austen's team collected this morning fit into the picture.

Austen leads me into a large, relatively cold, and dimly lit room adjacent to her office. This is the mesocosm lab where mud cored from the seabed and plopped into buckets is used to investigate the work of benthic animals. Rows of gray fiberglass tanks run almost the length of the room, each divided into four seawater-filled sections. The air temperature is kept at 59° F to help keep the water cool. Covers made of black landscape cloth draped over bamboo frames sit atop most of the tank sections or wells. All of this is designed for the comfort of creatures accustomed to the cold, dark subtidal seafloor. We have to raise our voices to be heard over the noise of the pumps moving saltwater through the tanks.

The newly collected Turritellas and other creatures the team pulled from the mud of the sound are sitting in black plastic holding tanks at the back of the lab now—except for the shrimp in their Teaboy cages, which are floating in one of the wells. Austen points out that a few of these creatures are on the shortlist of animals already known from past studies to be important for nutrient cycling. These include burrowing shrimps like *Upogebia* and *Calocaris*, heart urchins, brittle stars like *Amphiura*, small clams, and large worms such as polychaetes and sausage-shaped spoon worms (Echiurans). Even among these significant players, there are hierarchies of influence, Austen and her colleagues are finding. For instance, bulldozing heart urchins and burrowing shrimps have stronger impacts on nutrient cycling than Astarte and nut clams, brittle stars, or polychaetes such as *Nephtys*.[26]

Austen and Townsend position a rolling electric winch and begin lifting buckets from the morning coring into the wells one by one. The winch is necessary, Austen says, to prevent "benthos back," an injury caused by too much lifting of heavy buckets of mud. Townsend is using these and other cores to compare bioturbation in sand and mud sediments and a year later will be able to report that bioturbators have definite effects on nutrient cycling in these habitats, although as expected, the effects are stronger in mud than sand.[27]

On the floor at the back end of one long tank I notice smaller half-gallon buckets of sediment. Austen tells me these are filled with thousands of nematodes, protozoa, copepods, tardigrades, flat worms, ribbon worms, and other pinhead-sized animals. She imagines the sediment literally "fizzing" with the activity of a thousand of these tiny creatures, and she wonders how their accumulated impact might compare with that of 70–80 clams, shrimp, worms, or other large bioturbators.

Some nematodes and copepods do form burrows; others churn sediments as they feed or migrate up and down. Indeed, some marine scientists believe these tiny animals with their vast numbers may greatly influence the physical, chemical, and biological properties of sediments, including nutrient cycling, perhaps as much as the larger sediment animals in some habitats. Some of the larger animals actually serve in the ranks of "pinheads" during their juvenile stages. During that time, the teeming community of tiny animals just might help shape the assemblage of larger animals that share their habitat by preying on their larvae and influencing the living conditions the larvae must endure.[28]

Trawling, in general, seems to increase not only polychaete worms but also possibly the pinhead community.[29] Austen points out: "You get a shift to these smaller organisms, but whether they can actually substitute for all the ecosystem functions of the macrofauna and create a simplified system that can keep up nutrient cycling, we just have no idea."

She thinks that's unlikely, however, because tiny animals cannot stir up the deep sediments enough to let oxygen in and allow the nitrate wastes given off by microbial decomposers to diffuse into the water column.

"If you've got a diverse mix of bioturbators doing different things at different levels in the sediment—urchins bumbling around, worms moving up and down, bivalves at the bottom creating burrows, shrimps creating burrows—you end up with quite a deep oxygenated zone in the sediment," Austen points out. If the sediment harbors only tiny animals such as nematodes and copepods, the oxygenated layer will be much thinner. That means the nitrates produced during decomposition are more likely to end up in low-oxygen sediments where other bacteria will deactivate them [that is, denitrify the nitrate by converting it to biologically unusable nitrogen gas]. If less nitrate is released into the water, "then you've got less algal productivity and probably less fisheries," she says. "So in a nutshell, we think that probably a well-mixed and oxygenated sediment layer is likely to be the key thing, and to maintain that, we need a diversity of these bigger bioturbators."

Bioturbators perform other important work in the sediment besides stirring things up and indirectly servicing the food chain, and when they are reduced or eliminated by trawling, their loss can have ripple effects on the biodiversity of the seafloor.[30] Austen and Stephen Widdicombe, a senior scientist at Plymouth Marine Laboratory, have shown in a multiyear series of experiments at a mesocosm lab in Norway that some of these creatures exert a strong influence over which other species are present and in what numbers in their communities. Changes in the species of major bioturbators present—even shifts among species that are considered functionally similar in their feeding and sediment-disturbing habits—or changes in the density and distribution of bioturbators can alter the fate of other species in the community.[31] Dense patches of heart urchins scattered here and there through the sediment, for instance, can lead to greater diversity in nematode communities across a region.[32] The more heart urchins, brittle stars, or sea mouse polychaetes, the higher the biodiversity of other species in a community. The predatory polychaete *Nephtys* has the opposite effect—at high densities it lowers the diversity of other species in the neighborhood. In contrast, *Calocaris* shrimp, despite their burrowing, seem to exert no detectable impact on the diversity of other species in the community. All of the bioturbators just mentioned can be reduced or eliminated by trawling.[33]

Evidence from the fossil record suggests, in fact, that these "biological bulldozers" have exerted a powerful influence over sediment communities ever since they began to proliferate and diversify in the Devonian period 395 million years ago. Their rise has been credited with causing a parallel decline of immobile suspension feeders in soft sediment habitats. Many suspension feeders found in marine mud today are mobile burrow-forming species such as the mud shrimp *Upogebia* and the brittle star *Amphiura*. Rates of bioturbation and thus sediment stirring and disturbance are believed to have increased several orders of magnitude with the diversification of urchins, shrimps, lobsters, sea cucumbers, crabs, and other bulldozing animals. Some biologists believe the resulting acceleration in nutrient cycling helped to fuel greater algal productivity, diversification of algal species, and an explosion of new species of planktonic animals (zooplankton) that feed on algae.[34]

All these revelations about life in the mud may be exciting to ecologists, but how is a fisheries manager supposed to weigh the work of an urchin or worm or clam—or an entire fishing lane teeming with such beasts—against the social and economic benefits of a prosperous fishing industry? For that, Austen and her colleagues are turning to economists.

"If we say something is important, agencies want to know what to do about it," Austen says. "So we've got economists who are looking at these issues: If nutrient cycling changes, how do we put a value on that? If the biodiversity changes, can we value that? We are trying to set up a decision-support system for management to help them weigh up the pros and cons and balance biodiversity against economic outcomes, jobs gained, et cetera, by adversely affecting the benthos. What happens if you limit fishing or gear types in certain areas or declare marine protected areas? That's where this could and should go."

Attempts to value the ecological services a healthy ocean provides have been few and have focused at a global scale. In 1997, a team of ecologists and economists led by Robert Costanza, now director of the Gund Institute for Ecological Economics at the University of Vermont, created a stir when they estimated the minimum value of the world's ecosystem services and "the natural capital stocks that produce them" at US$33 trillion per year. (Natural capital includes trees, animals, rivers, oceans, and other natural resources and systems that generate goods and services vital to human welfare, including the creation of manufactured capital such as machines and buildings.) The global oceans account for two-thirds of that value because of their vital role in regulating climate and the cycling of water, nutrients, and carbon; absorbing and diluting contaminants; and providing us with food, recreation, and employment.[35] What portion of that value is supplied by the mud-dwellers of the global seas? The mud-dwellers of a regional fishing ground? How is that value affected by various fishing practices?

Although economists and marine scientists grapple with answers that will fit into cost-benefit calculations, fisheries managers are beginning also to acknowledge the intrinsic value of some seafloor habitats. Since the 1990s, for instance, the extent and intensity of bottom trawling in U.S. waters have been reduced as managers closed some areas to fishing, restricted fishing seasons, or required gear modifications to minimize bottom contact, usually in response to declining fish stocks. A prime example is the ban on trawling in the wake of a collapse in cod and haddock stocks in New England and off Georges Bank (east of Massachusetts).[36]

At the urging of ecologists and environmental groups, some unique benthic communities such as slow-growing deepwater corals—most of them only now being explored and charted—have gained protection from trawling in a few areas: the Oculina Banks off Florida's Atlantic coast, some coral beds off Nova Scotia, 19 seamounts in New Zealand waters, and an extensive reef area off Norway.[37] Just that morning, in fact, we had heard on BBC radio that the EU would issue an emergency ruling closing the coral-rich Darwin Mounds off the northwest coast of Scotland to trawling for 6 months while European nations considered a permanent closure.

Soft-bottom marine life is getting little, if any, protection right now in most of the world, and few environmental advocates champion its cause. Yet in Austen's view, the vast subtidal mud plains are to ocean life what the Serengeti plains are to African wildlife.

"The analogy I've often used is this: If you go to the Serengeti plain and look at all the mammals that are there—zebra, wildebeest, lions—they're all completely dependent on the grass. And if the grass were plowed up on a regular basis, the rest of the stuff on top would disappear sooner or later. I think we're talking about the same sort of habitat here. If you just keep plowing up the seabed, you're eventually going to lose the life in the waters above it, one way or another."

It occurs to me, however, that as fish stocks collapse, we might not be able to tease out these bottom-up effects on the food chain that Austen is talking about from the impacts of overharvesting, ocean warming, pollution, and all the other direct human affronts to the oceans. By 2002, for instance, 72 percent of the world's ocean fish stocks were being harvested faster than they could reproduce.[38]

"It's a bit of a race, isn't it, whether they just scoop up so many fish that that has a quicker effect than what they're doing to the seabed," Austen acknowledges. "They're undermining it from two levels, top-down and bottom-up. We're just looking at how important is the bottom-up effect at the moment, and there are plenty of fisheries people who are looking at how important is the top-down cascade effect of taking out the top predators, then the ones below, then the ones below them. You're hitting it both ways, and neither can be terribly good long term. You can't make a sustainable fishery like that.

"If you went into the Serengeti and just plowed up bits of it occasionally in a controlled way, then you could probably keep it so that you'd have just enough habitat to maintain viable populations. But there's no way you can keep plowing up more and more of the seabed and maintain populations; and there's no way you can keep harvesting more and more fish and maintain viable populations forever."

VI

Microbes, Muck, and Dead Zones

A few years ago, someone stenciled the sidewalks near my Montana home with footprint-sized silhouettes of trout. The symbols appeared at street corners near the openings of storm drains. The message "Dump no waste, drains to river" bracketed each fish symbol. The habitat of primary concern to the painters was the Gallatin River, a near-pristine trout stream that tumbles north out of Yellowstone National Park to form one of the three headwaters of the Missouri River. Lawn chemicals, road oil, industrial fluids, and sewage discharges from Bozeman and other towns in the Gallatin Valley, along with manure and fertilizer from hay, wheat, and potato fields, all seep into the passing river and its feeder streams.

The Gallatin is only the first stop, however. Our sidewalks might just as legitimately bear the silhouette of a red snapper or brown shrimp, I've recently come to recognize. Whatever contaminants wash off the lawns here or nearby crop fields into the Gallatin can eventually travel more than 3,700 river miles down the Missouri into the Mississippi and end up flushing into the northern Gulf of Mexico.

In truth, farmers and city dwellers in 30 other U.S. states should also be thinking about the gulf's snapper and shrimp. The Mississippi

River watershed (all the land that contributes water to the river) sprawls across more than 40 percent of the land in the lower 48 states, including more than half the country's farmland. It receives wastes discharged from sewage plants and other specific effluent pipes or "point sources" of pollution as well as millions of dispersed or "non-point sources," such as cornfields, cattle pens, and golf courses.

All this material, but particularly the nitrates (NO_3) leached from fertilized cornfields in the upper Midwest, combines to degrade water quality and habitat in rivers across the watershed. Far downstream, out of sight and mind of those of us who generate it, 1.6 million tons of that nitrogen spill into the coastal waters of the gulf each year. That triggers a complex cascade of events that starts when the nitrogen—an essential plant nutrient—fertilizes excessive blooms of small floating algae or phytoplankton. The algal blooms in turn spur a population explosion among microscopic grazing animals known as zooplankton. When this plankton productivity is too great for fish and the rest of the food chain to use—a state called eutrophication—uneaten organic matter sinks to the seafloor to rot, prompting a population boom among decomposer microbes. The frenzy of heavy breathing by these microbes, in turn, uses up dissolved oxygen. Oxygen in the air is readily soluble in water, and surface water in contact with air is rich in dissolved oxygen that most aquatic organisms require to power their metabolism. In seasons when there is little mixing between the surface and deeper waters and biological activity is high, however, the oxygen in bottom waters becomes depleted. The result is a recurring summer "dead zone," a vast area of oxygen-starved, or hypoxic, bottom waters starting near the mouth of the Mississippi and Atchafalaya rivers and reaching far into the gulf that drives fish, shrimp, and other mobile creatures from an area larger than Lake Ontario—an area that reached a record 8,000 square miles in July 2001.[1] Clams, oysters, sea stars, and many other benthic creatures unable to travel far or fast enough to escape simply suffocate.

The problem of oxygen-starved waters is not confined to the Gulf of Mexico, although the gulf hosts the second largest hypoxic zone in the world. More than 60 percent of coastal rivers, bays, and estuaries in the United States are moderately to severely degraded in this way by excess nutrients. The condition is particularly acute in Long Island Sound, Chesapeake Bay, and the Florida Keys.[2] Worldwide, the United Nations Environment Programme recently identified nearly 150 hypoxic zones in coastal waters, all of them linked to excess plant nutrients, mainly nitrogen, from varying combinations of inefficient or excessive fertilizer use, discharges of untreated sewage, and airborne emissions from vehicles and factories. All three of these nitrogen sources plague the Baltic Sea, which hosts a dead zone more than triple the size of the one in the Gulf of Mexico. Other dead zones can be found in the Black Sea, the northern Adriatic, the Kattegat Strait between Denmark and Sweden, and in waters off South America, China, Japan, Australia, and New Zealand. Overfertilized waters are expected to emerge soon off the coasts of Asia, Africa, and Latin America with continued industrialization and more intensive agricultural production.[3]

Troubling as these dead zones are, the problems caused by excess nutrients would be much worse without the natural "life support services" provided by rooted plants and especially by bacteria in the mud of waterways, wetlands, and floodplain forests throughout the world's watersheds. Thanks largely to the natural cleansing services of these organisms, for example, only one-fourth of the Mississippi River's increasing burden of waste nitrogen actually reaches coastal waters.[4] Clearly, however, in the Mississippi basin and many other watersheds, the scale of human activities has begun to overwhelm the capacity of sediment communities to use, recycle, or eliminate reactive (biologically usable) nitrogen.

The fact is, human activities have more than doubled the amount of nitrogen that would be circulating through the earth's systems thanks to natural inputs alone. Each year we put another 160 million metric tons of reactive nitrogen into play through fertilizer production and use, fossil fuel burning, and planting of legume crops such as soybeans that use nitrogen captured ("fixed") from the air by symbiotic microbes. That compares to 90–120 million tons of nitrogen drawn from the air by similar nitrogen-fixing microbes in natural set-

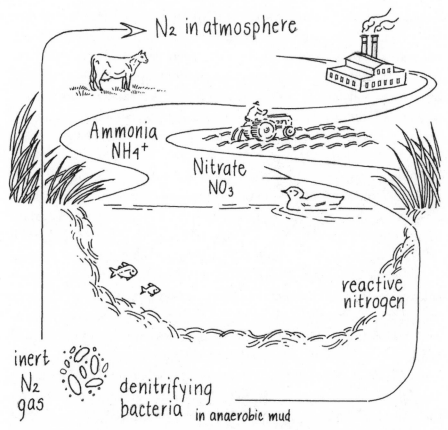

Wetland sediments host vast populations of denitrifying bacteria that can inactivate and release to the atmosphere much of the excess nitrogen generated by our farms, livestock operations, factories, and automobiles.

tings (and a bit more by lightning). The higher the human population density in a watershed and the greater the quantity of nitrogen put into play, the greater the nitrogen load in our rivers.[5]

What humans haven't increased is nature's capacity to turn that nitrogen back to the atmosphere in its inert (biologically unusable) form, a recycling job that a rich array of microbes handle through a process called "denitrification." In fact, we've often severely reduced denitrification services in regions such as the Mississippi River basin, where we need them most, by destroying wetlands that serve as home to vast concentrations of denitrifying microbes. Admittedly, that extra nitrogen we generate in the form of fertilizer and through burning coal and oil makes it possible to feed and fuel our burgeoning populations, but it also helps drive some of our most severe environmental problems, from acid rain, global warming, ozone depletion, and loss of plant diversity to polluted water supplies and coastal dead zones.[6] The attempt to redress some of this damage by restoring prime habitat for denitrifying microbes in the Mississippi basin is the topic of this chapter.

The Olentangy River lies about 1,500 miles east of the Gallatin, yet its waters end up in the same place. The river flows through Columbus, Ohio, on its way to the Ohio River, then the Mississippi, and on to the Gulf of Mexico. On a sunny late afternoon in May, I drive north along the river through the Ohio State University campus and arrive at the Olentangy River Wetland Research Park just as the staff is leaving for the day. When I ask for the director, William J. Mitsch, his assistant sends me down a paved path toward a two-story pavilion overlooking a series of manmade marshes. Mitsch, an unabashed evangelist for the virtues of wetlands, has just set off on a tour of the 30-acre site with two dozen local science teachers in tow. I catch up and fall in behind the group just as Mitsch is pointing out a sign on the ramp leading up to the pavilion. It touts the three great values of wetlands: flood control, water purification, and wildlife habitat.

"We've got to start thinking of the floodplain as part of the river," Mitsch tells the teachers as they crowd onto the ramp. "We calculated 6 million gallons of water were here in the wetlands and not in somebody's basement during the January floods." The wetlands he is talking about stretch out before us: a 7-acre "billabong," an Australian term for the oxbow wetland that curves along the bordering Olentangy River, and a pair of kidney-shaped experimental marshes, each covering 2 1/2 acres.

The kidney shape was a symbolic choice for Mitsch. Wetlands not only serve as a relief valve for floodwaters but also function as the river's "kidneys," filtering and cleansing the water that passes through them. Today the river is running high and fast, and the pumps that control water intake in these marshes are pulsing the water in quickly to mimic natural seasonal flooding. A drop of water now spends only two days in residence in the marshes instead of the usual three or four. "A lot of biology and chemistry goes on during that time," Mitsch tells us. "When water moves through the wetland and out more quickly, it doesn't have as much time for contact with the sediment, and soil-water exchange is very important for things like nitrate removal. So what we'll see in our samples is a smaller percentage of nitrate removed during the floods." Each year, though, these little marshes remove about 3,500 pounds of nitrogen and 27 tons of sediment from the Olentangy before returning its waters to the river on their journey to the gulf.

The habitat value of wetlands is on active display around us as swallows dart low across the water and red-winged blackbirds flit about in the reeds. A muskrat cuts a V-shaped wake across the billabong as it swims. From the shaded upper deck of the pavilion, we can see a great blue heron stalking frogs and small green sunfish. What we cannot see, though, are the microscopic creatures in the ooze that perform most of the real work of water purification for which wetlands receive credit.

Unfortunately, none of the ecological values of wetlands—to say nothing of their value for joggers, walkers, and birdwatchers—were widely recognized or appreciated until recent decades. For most of human history, swamps have been regarded as miasmic, mosquito-ridden wastelands, and any money and effort directed their way largely went toward draining and "reclaiming" them. As a consequence, more than 80 percent of the wetlands in Ohio and other states in the heart of the Mississippi River basin have been drained, often to create more farmland that in turn leaks more nitrates.[7] What's more, extensive subsurface drainage networks installed in previously waterlogged farm fields throughout the basin allow nitrate-laden water to bypass remaining wetland or forest buffers and flow directly into lakes and streams. And opportunities for rivers to spill out across floodplains and deposit nutrient-rich sediments back on the land before reaching the sea have also been vastly diminished. Extensive construction of levees and flood-control works in the first half of the

20th century, for example, effectively isolated the Mississippi River from its floodplain.[8]

Mitsch leads us to a footbridge over the Olentangy at the edge of the wetland complex, then downriver atop a 6-foot-high earthen levee. This levee was erected a century ago to keep what is now the experimental wetland from flooding and allow the land to be farmed. It also blocked periodic flooding of the bottomland hardwood forest that fringes the river. "These streamside forests that periodically flood were the most common natural wetlands in Ohio, and they can have trees that are up to 100 years old," Mitsch says, coming to a halt at a small channel cut through the levee. Chocolate-colored water is flowing through, and the trees beyond—all adapted to periodic "wet feet"—are standing in a foot of water.

Four years earlier, Mitsch had convinced the state highway department—under legal requirement to atone for its destruction of wetlands on a road project—to cut four "notches" in this levee. "What we got was a quadruple bypass for this forest that restored its circulation and the integrity of its function," he tells us. The 13-acre forest is dry most of the time, but 4–10 times a year, floodwaters like these surge through the breaches, depositing vital nutrients that would otherwise be carried downstream.

Like marshes, bottomland forests in the Mississippi basin disappeared rapidly during the past century, many as a result of federal flood-control projects. In 1937, half the Mississippi floodplain was forested. By 1977, less than one-fourth was forested, and clearing and draining continue even now.[9] Since 1988, rampant destruction of wetlands in the United States has been dampened by a "no net loss" of wetlands policy and requirements to replace the acreage destroyed, but controversy continues about the effectiveness and enforcement of the policy. The very existence of the policy, however, does indicate how much society's view of the multiple values of wetlands has changed. Mitsch entered the field of ecology just in time to be swept up in that new appreciation of wetlands.

Mitsch began his career as a mechanical engineer but was quickly caught up in the wave of environmental consciousness that led to the first Earth Day in 1971. He headed to the University of Florida, Gainesville, to study with Howard T. Odum, a pioneering systems ecologist who was then launching research on the use of wetlands for recycling municipal wastewater. It was Howard Odum and his brother, ecologist Eugene P. Odum, who first called the vital, self-sustaining natural ecosystems of the earth our "life-support systems."[10]

Training in engineering and ecology left him with a hybrid mindset, Mitsch says. He retained the problem-solving, whole-systems outlook of an engineer, yet turned to the ecology of muck, microbes, and reeds to find solutions to problems that engineers would tackle with scrubbers, filters, and fossil fuel–driven processes. He has spent three decades developing a field Howard Odum initiated under the name "ecological engineering," a field committed to solving environmental problems by restoring the essential "bodily functions" of Mother Nature.[11] Some of her most vital organs, of course, are wetlands. And it was his research on restoring and creating wetlands that got Mitsch involved with the problem of hypoxia in the Gulf of Mexico and led him and a handful of colleagues to advocate restoration on a massive scale.

Since the early 1970s, research teams surveying groundfish or exploring for oil and gas in the northern Gulf of Mexico have reported finding patches of oxygen-depleted bottom water. Not until 1985, however, did anyone begin systematically mapping coastal waters to determine the extent of what has come to be called the "dead zone." A team led by marine ecologist Nancy Rabalais of the Louisiana Universities Marine Consortium in Cocodrie has mapped the phenomenon every summer since then, sampling the dissolved oxygen content of the seawater at more than 60 sites. In 1991, with 5 years of data in hand, Rabalais and her team revealed for the first time the extent of the dead zone and confirmed that it returned every year. Then in 1993, record floods throughout the Mississippi River basin sent vast quantities of nitrogen-laden waters into the gulf, doubling the size of that summer's dead zone and catapulting the issue to national prominence.[12]

There have always been naturally occurring hypoxic zones in the seas, including deep-ocean "oxygen-minimum zones," but overfertilization with nutrients is increasing their occurrence in estuaries and shallow coastal waters.[13] Coastal systems such as the northern gulf are especially vulnerable because their waters are naturally stratified in spring and summer, preventing mixing between oxygen-rich surface waters and isolated bottom waters. Stratification occurs in spring when large plumes of relatively warm, buoyant freshwater generated by upstream rain and snowmelt pour out across the surface of the gulf, forming a virtual lid that traps the cooler, denser seawater below it. By fall, storm winds and cold fronts whip up the waters and end both the stratification and the hypoxia. In summer 2003, two tropical storms occurred during July, roiling the gulf and reducing the hypoxic zone during the mapping period to less than half of average for the decade. Relying on such occasional summer storms to break up the dead zone, however, is hardly a viable solution. Although seasonal stratification is natural, hypoxia in the gulf is not. Studies of sediment cores, in fact, reveal that the severity of oxygen stress in the gulf has been increasing dramatically since the 1950s.[14]

The definition of hypoxia for the northern gulf is less than or equal to 2 parts per million dissolved oxygen, less than half the normal oxygen levels for those bottom waters and a level below which trawlers don't catch shrimp and bottom-dwelling fish.[15] Trawlers now compensate by moving farther offshore to fish in summer, and no decline has been reported in the northern gulf fishery, which accounts for one-fourth of the U.S. commercial fish catch. But marine biologists consider the growing dead zone a potential "time bomb" for both biodiversity and fisheries.[16] In other oxygen-stressed coastal waters such as the Kattegat and the Black Sea, damage to fisheries and marine ecosystems has grown progressively worse as oxygen stress intensified.[17]

Biologists are concerned, for example, about changes occurring in both the benthos and the plankton communities at the base of the food web. Recurring hypoxic episodes tend to shift the makeup of the seafloor community away from large, long-lived species such as clams and sea stars to smaller, short-lived creatures such as polychaete worms whose larvae can recolonize the zone quickly between hypoxic episodes.[18] Loss of suspension-feeding and bioturbating species in addition is likely to intensify the eutrophication process.[19] Robust populations of filter-feeding mussels, oysters, scallops, or clams, for instance, can remove a great deal of excess algae from the water. The American oysters of Chesapeake Bay once effectively filtered the equivalent of the entire volume of the bay every 3 days. The mismanagement and collapse of these oyster populations have increased the eutrophication rate in the bay.[20] Loss of the bulldozing and sediment-aerating services of deposit-feeders, whether to hypoxia or to trawling and dredging, can also interfere with nutrient cycling and enhance eutrophication.[21] All of these direct and indirect effects of hypoxia can be compounded and perhaps masked by the impact of other pressures such as overfishing and offshore oil and gas production.[22]

In 1995, environmental and fishing groups invoked the powers of the U.S. Clean Water Act and petitioned state and federal officials to take action on the gulf dead zone. That set in motion an "integrated assessment" by some 50 scientific experts on six teams, one led by Mitsch, of the extent of gulf hypoxia, its causes, the ecological and economic impacts, and possible solutions.[23] As to cause, the technical assessment pointed directly to the burgeoning levels of nitrogen coming down the Mississippi, noting that the annual nitrogen influx to the gulf has tripled since the 1950s, paralleling the increase in hypoxia. The majority of that nitrogen arrives in the form of nitrates washed off of Midwest farms.[24]

Fertilizer is the largest source of nitrogen entering all the rivers that supply freshwater to the North Atlantic Ocean, not just the Gulf of Mexico. On average, less than half of the nitrogen fertilizer applied to fields worldwide gets harvested as crops. Much of the rest is dissolved and washed directly into waters or volatilized to the air in forms that can rain back to earth.[25] Even the nitrogen taken up by crops and incorporated into greenery and grain can eventually end up in waterways in the form of human or animal wastes. About 40 percent of the global grain harvest is fed to livestock, which generate nitrogen-rich manures that can wash into streams or release ammonia to the air.[26]

Clean water laws have spurred progress in some parts of the world in cleaning up obvious point sources of pollution such as industrial effluent pipes that discharge into lakes and rivers. But few nations have tackled the growing problem of pollutants such as manure and fertilizers that leak from the landscape. In the United States and most of the rest of the world, we still rely by default on aquatic systems to make these nonpoint contaminants "go away."

Once the technical assessment was completed, a task force of federal, state, and tribal government agencies from throughout the Mississippi basin drew up an "action plan" for reducing and mitigating hypoxia in the gulf. The 2001 plan that resulted represents a compromise among many competing interests, from fishermen to powerful farm groups. The task force set a target of reducing the average extent of the dead zone to 2,000 square miles by 2015. To accomplish that, the plan assumes that nitrogen discharges to the gulf will need to be reduced by 30 percent. That reduction would be accomplished for the most part by a two-pronged strategy: reductions in nitrogen entering the river system and extensive restoration and creation of wetlands—mostly in the upper part of the basin—to trap nitrogen that does end up in the river.[27] The upper basin states would thereby benefit as well as the gulf in this scheme. The U.S. Environmental Protection Agency considers at least 44 percent of rivers in the basin states to be "impaired"—with, for example, nitrate concentrations near or above the statutory maximum for drinking water (10 parts per million) and a history of pollution-related warnings about eating fish.[28] Restored wetlands and better fertilizer management in the Midwest would not only improve water quality but also provide other values mentioned above: flood relief, wildlife habitat, and recreation.

The enormous changes needed in farm practices, land use, and watershed restoration are to be accomplished by incentive-based voluntary actions in each region rather than by regulation. So far, the task force has done little to coordinate any basinwide efforts or secure new sources of funding. But that has not kept some state and regional groups and researchers from moving ahead. The assessment team Mitsch led, for instance, estimated in 1999 that nitrogen reaching the gulf could be reduced by as much as 40 percent through restoration and creation of an unprecedented 5–13 million acres of wetlands and 19 million acres of floodplain forests. The team also recommended creating more flood diversion structures in Louisiana so that flood-waters can be shunted through an extra 1–2.5 million acres of back-waters and wetlands in the Mississippi River delta before reaching the gulf.[29] In the years since, Mitsch and assessment team colleague John Day, a professor of coastal ecology at Louisiana State University in Baton Rouge, have led a continuing effort to refine those numbers based on findings about the nitrogen-processing capacity of wetlands and their microbial inhabitants.

The hallmark of wetlands, as their name implies, is that their soils are saturated or inundated with shallow water.[30] As plant roots, microbes, and other respiring creatures draw oxygen from the sediments, they eventually use up the oxygen and thereby render the sediments anaer-obic. Because of this, wetland soils support a wide range of anaerobic microbial processes such as fermentation, which produces alcohol and acetic acid, methanogenesis, which produces methane, sulfate reduc-tion, which releases hydrogen sulfide, and denitrification, which releases nitrogen gas into the air. The microbes involved in these processes use creative ways to obtain energy as they dine on and decompose organic matter in the absence of oxygen. (Indeed, the metabolic versatility of microbes is one reason that "dead zones" and other extreme envi-ronments are never truly devoid of life.) The wide range of chemical expertise and appetites of microbes resident in wetlands is responsi-ble for the cleansing power of these ecosystems. The nature and work of one group—the denitrifiers—are of particular importance, as the name implies, in coping with excess nitrogen.

Nitrogen is an essential building block of organic molecules such as chlorophyll, DNA, proteins, and the enzymes that catalyze life processes. The air around us is 78 percent nitrogen and 21 percent oxygen, yet we cannot use the nitrogen we breathe the way we absorb oxygen. Few creatures can. That's because nitrogen molecules in the air are tightly paired into "di-nitrogen" or N_2 form, rendering them inert and unavailable for use in metabolic reactions. A few soil microbes, however, have developed the ability to split that bond and "fix" N_2 into usable forms. Some symbiotic microbes can accomplish this energy-intensive task, including bacteria of the genus *Rhizobium*, which partner with the roots of legumes, and actinomycetes (bacteria that grow in threadlike form similar to that of fungal hyphae) in the genus *Frankia*, which team up with a variety of other plants. Some free-living microbes such as cyanobacteria and cyst-forming *Azotobacter* also fix nitrogen. That was it until 1913, when an industrial nitrogen-fixing process known as Haber-Bosch was invented and humans began to pull ever increasing amounts of reactive nitrogen from the air. The first use of this new process was to increase supplies of nitrate for manufacturing military explosives, but since the 1950s, most of the reactive nitrogen produced has gone into making synthetic fertilizers.

Once biologically reactive nitrogen is put into play, an array of microbes determines its fate. Some microbes break down organic forms such as the amino acids in proteins and the nucleic acids in DNA to yield ammonia; nitrifying bacteria convert ammonia to the more plant-usable form nitrate; ammonifiers reverse the process, converting nitrate to ammonia; and denitrifiers convert nitrate back to gaseous forms—mostly to inert N_2 but some fraction is converted to nitrous oxide (N_2O), a culprit in global warming, destruction of the stratospheric ozone layer, smog formation, and acid rain.[31] Only when some stages in this complex, interrelated set of nitrogen transformations are reduced or overwhelmed do we notice—such as when excess nitrate pours into the gulf and causes hypoxia.

For those interested simply in the large-scale cleansing capacities of wetlands, denitrification is a black-box process to be measured in terms of how much nitrogen flows into and out of a system. Few are familiar with the identity, preferences, and capabilities of the one-celled creatures that actually do the work. To learn a bit about these creatures, I turned to James M. Tiedje, director of the Center for Microbial Ecology at Michigan State University, who has long studied the ecology of denitrifying bacteria and how denitrification is regulated in nature. The first thing I learned is that denitrification is a job that a very broad array of microbes can handle, and so denitrifiers are probably ubiquitous in soil.

"It's easier to say which kinds of groups don't have denitrifiers among them," Tiedje told me when I asked for some examples. That would include gut-dwelling organisms such as *Escherichia coli* (*E. coli*) and also groups like fermenters and methanogens that must operate without oxygen. "But among most other major physiologic groups, there are some denitrifiers," he added. Even certain nitrogen-fixing bacteria can undo their work by denitrification. "Probably most of the denitrification in soils would be carried out by some of the most common, rapidly growing soil organisms, like *Pseudomonas*." *Pseudomonas* is a large and versatile genus whose members engage in a wide range of activities including decay and nutrient cycling, promoting plant growth and disease resistance, or causing disease among plants, animals, and humans. Many pseudomonads are also nutritional opportunists, known for getting their carbon by eating toxic pollutants such as fuel oil. And some pseudomonads can also denitrify nitrate when they must.

I say "when they must" because it turns out that no microbe lives the life of a denitrifier when it has a choice. Virtually all denitrifiers prefer to respire using oxygen just like we do, Tiedje explained. But when oxygen becomes scarce, denitrifiers can feed on nitrate to obtain the oxygen molecules they need to "burn" (respire) carbon during metabolism. They then exhale nitrogen gas as a waste product. Soil bacteria can find themselves forced to make do without oxygen not only in the flooded soils of wetlands or ocean sediments but also within tiny soil aggregates almost anywhere from forests and crop fields to compost heaps.

Denitrification is a second-choice lifestyle because the payoff is meager compared to breathing oxygen. "The energy yield from denitrification compared to respiration with oxygen is about two-thirds," Tiedje told me. "So that's why it's to the organisms' benefit to always use oxygen when they can, because they make one-third more cells."

There are several oxygen-containing (oxidized) inorganic compounds besides nitrate that can be respired, including sulfate, ferric iron, and some chlorinated compounds. "They all yield different amounts of energy, and nitrate is one of the best," he said. "So if you can't do oxygen but you can do nitrate, you usually have a competitive advantage over everybody else." And all of the alternative ways of respiring are more efficient than having to ferment organic compounds such as glucose to get energy.

Do different types of microbes handle the same job differently, and does it matter? This is a hot topic of research right now, for denitrification and other microbial processes. Some bacteria, for instance, might be particularly good at denitrifying in the presence of extremely low levels of nitrate. Others might produce and release the undesirable byproduct nitrous oxide more readily. One difficulty in probing such questions has been identifying just who is doing the work. For example, pseudomonads—or what researchers thought were pseudomonads—have been used extensively in research on the biochemistry of denitrification.

"Twenty years ago, *Pseudomonas* as a genus was huge," Tiedje recounted. "Then beginning in the 1990s, it was split into primarily five other genera." Even to a non-taxonomist, that indicates these bacteria may not be such close neighbors on the family tree. It also means that some of the activity that was attributed to pseudomonads in various experiments was actually the work of other genera of denitrifiers.

Even with molecular techniques available, researchers are finding it difficult to parse microbes into what biologists dealing with larger life forms call "species." The entire genetic blueprints for 150 microbes have been decoded, for instance, including complete genomes of nine strains of the species *E. coli*. Yet only about one-third of the *E. coli* genes are common to all nine. "That's a lot of functional diversity that can occur in what is currently considered a single species," Tiedje pointed out. "Generally, what we want to use the species concept for is to be somewhat predictive of what an organism is capable of doing. So the current species definition is too liberal."

Many people dealing with the ecology of microbial processes simply ignore species and stick to the black-box level, but Tiedje thinks it's time for a change. "I'm of the opinion that we've looked at these general processes for 30 years and now we need to know more about the organisms underlying those processes, the catalysts," he said. That's especially true for microbes that could be harnessed to clean up environmental pollutants. Finding a low-concentration specialist that eats benzene, for instance, is important when you want to clean up the last bit of benzene residue in drinking water. Some organisms that eat benzene, toluene, and other serious contaminants in gasoline are also capable of respiring nitrate, Tiedje noted, a combination of talents that could be put to work cleaning up oxygenless underground pollution sites.

For removing dissolved nitrate in wetlands, however, Tiedje believes the diversity of organisms involved and their wide distribution make it reasonable to ignore the particular identity of the microscopic players and manage the process with whatever microbes are present. Studies by Tiedje and other researchers have identified a number of environmental conditions that affect the overall rate and efficiency of the denitrification process—probably by influencing the performance of various organisms and the structure of the microbial community in ways scientists haven't yet deciphered. Warmer waters, low salinity, and plenty of edible carbon boost denitrification, for instance. Heavy metals such as cadmium, copper, and zinc can inhibit it. Fast movement of water through a wetlands system reduces the percentage of nitrate that can be removed.

Research on understanding the process continues. Even as Mitsch and I sit in his conference room talking, we can see that one of his students has donned knee-high rubber boots and sloshed out into the flooded margins of the marsh to draw air samples that will eventually tell her whether more nitrogen is being released to the air from permanently flooded sediments or from soils that are intermittently wet and dry.

A fundamental tenet of ecological engineering is whenever possible to rely upon the "self-designing capability of ecosystems and nature."[32] Thus Mitsch sees little need to manipulate the plants, much less the microbial community, in most created wetlands, as long as designers get the water circulation right and are willing to be patient.[33] Patience is sometimes needed for decades, not the 5-year time frame by which most constructed wetlands are judged a success or—more often—a failure. His is a "build it and they will come" approach that not all ecologists agree with, to be sure.[34] Indeed, federal regulations for constructing replacement wetlands often specify which plants must be present in what proportions. But unless the goal is to restore habitat for a specific plant or animal, Mitsch believes that's unnecessary.

"Get the hydrology right and you'll get an abundance of diversity coming in out of the river," he says. Bacteria, algae, insects, fish, and plant propagules of all sorts will wash in. Seeds will arrive on the wind. And Mother Nature will choose what survives.

"I think you can tell after a couple years whether you've got the water right, and if you get that right, Mother Nature will be very, very forgiving and eventually it will develop into a very nice wetland," he tells me as we sit looking out on the marshes. "I just believe that. It will go through some pains. It'll have cattails invading, muskrats coming in. But I think in the long run, if it's a sustainable hydrologic situation, then you're well on your way to being successful."

To test this approach, Mitsch and his team planted one of the newly built kidney-shaped marshes in 1994 and left the second one for "Mother Nature" to plant. A decade later, neither the teachers nor I succeeded when Mitsch challenged us to pick which was which. The marsh just below the pavilion, it turned out, had been planted with native trees and bulrushes. Nature favored cattails at first on her side, thanks to the nutrient-rich soil that had been farmed for decades. Now a line of cottonwoods and willows, some 10 feet tall, are edging into nature's side of the marsh, all sprung from seeds blown in from the riverbank forest. Hundreds of other plants, insects, birds, and other creatures have found their way into the marshes, too. There are still differences: more fish in the planted side, more turtles and frogs in nature's half, for example. But in appearance and functioning—especially in the nitrate-removal capacity of their sediment communities—the two are converging.[35]

The fate of these decade-old wetlands has implications for the gulf because the scale of restoration Mitsch and his colleagues believe is needed in the Mississippi River basin will preclude micromanagement. Further, these wetlands experiments are not only testing ideas about self-organization but also helping the researchers fine-tune their recommendations about the extent of new wetlands that will be required to staunch the bleeding of nitrogen from the heartland to the gulf.

As a general rule, the amount of nitrogen that runs off the land and into rivers drops as the percent of the landscape covered by marsh increases.[36] There is a point of diminishing returns, however, after which, if you spread the nitrogen-laden water across more acres of wetlands, the amount of nitrogen removed per acre simply declines.[37] Thus Mitsch and others ponder the optimal proportion of the landscape to devote to wetlands, for water quality, flood control, and other values.[38]

Before European settlement, about 10 percent of the landscape in the Upper Mississippi and Missouri river basins was probably covered with wetlands. Around 2 percent is wetland today. After the disastrous flood of 1993, one research team calculated that restoring enough wetlands to bring the proportion back up to 7 percent would provide adequate floodwater storage to prevent a recurrence.[39] To reduce nitrogen loads in the Mississippi by 20–40 percent, the assessment team led by Mitsch originally estimated that another 1–2 percent of the Mississippi basin would need to be converted back to wetlands and 3–7 percent to riparian forest to provide enough habitat for the new armies of denitrifying bacteria that would be needed.[40]

To refine these estimates, Mitsch and John Day of Louisiana State have continued to collect information on the amount of nitrogen removed per acre in a wide variety of wetland types, from the little Olentangy research marshes to the gigantic Caernarvon diversion system south of New Orleans, which since 1991 has been channeling water from the Mississippi main stem into a 100-square-mile wetland as part of an effort to restore deteriorating marshes in the delta.[41] Although the nitrogen-trapping performance of any given wetland varies over time—with the seasons, with changes in nitrate levels in the river, and with a host of other environmental factors—the team has now accumulated a combination of 40 years of data from a variety of marshes in Ohio, Illinois, and Louisiana. A dozen years of monitoring here at Olentangy, for instance, show that on average the experimental marshes pull 498 pounds of nitrate per acre from the water each year. Louisiana's Caernarvon wetland performed at a slightly higher average rate over 12 years of sampling, with each acre of marsh removing 586 pounds of nitrate from Mississippi River water.

Mitsch, Day, and their colleagues combined these "retention rates" with data from another 26 years of studies at other sites in the basin to develop a model of the average amount of nitrogen removed per acre when rivers carry levels of nitrate that are now typical of the Mississippi River watershed. The model yielded an estimate that 5.4 million acres of new wetlands (0.73 percent of the land in the basin and an area more than four times the size of Delaware) will be needed to reduce the nitrogen flowing into the gulf by 40 percent. Although the number falls at the lower bound of their previous estimates, it's by no means business as usual for wetland restoration and creation. That 5.4 million acres represents 65 times more wetlands than the United States has netted nationwide over the past decade through enforcement of clean water laws, and four times the amount protected or restored nationwide under the U.S. Department of Agriculture's Wetland Reserve Program.[42]

It's also a bigger number than you'd get from an environmental engineer (formerly called a sanitary engineer) constructing wetlands for tertiary treatment of wastewater (the nutrient removal stage) from municipal sewage plants, a wetland service that has been well studied since Howard Odum's pioneering work.[43] "We're the only ones coming up with these gigantic numbers for wetlands that would have to be restored," Mitsch tells me. "People who design wetland waste treatment systems will tell you they can remove five or six times more nitrogen per acre than we're getting, and they can. But treatment marshes built as part of a sewage plant are not meant to be sustainable, and they inevitably become cattail marshes. We're talking about building wetlands that are functional, sustainable, have some biodiversity and all the other values that wetlands can provide. We don't want single-purpose systems, especially up here in the upper Midwest just to benefit the gulf."

It's an example of the fact that trying to maximize a single ecological service, whether water purification or crop and timber production, often creates "disservices" in other ecological functions.

Some worry, in fact, that millions of acres of wetlands with denitrifying microbes at work will mean much greater amounts of the greenhouse gas nitrous oxide being emitted to the atmosphere, even though it represents only a small fraction of the gas the microbes release. Mitsch's assessment team concluded, however, that the amount of reactive nitrogen that humans put into play ultimately determines how much denitrification takes place and thus how much nitrous oxide is released. Without wetlands to do the work, much of that denitrification is probably occurring in gulf sediments after the nitrogen has already done its harm in triggering hypoxia.[44] Ultimately, the only way to reduce nitrous oxide releases is to reduce nitrogen use. So along with wetland creation, Mitsch and Day are still calling for better farm management practices to reduce nitrates entering the river system by 20 percent.[45]

The 5.4-million-acre figure of new wetlands needed to reduce by 40 percent the nitrogen entering the gulf does not include bottomland forests because these usually remove less nitrogen per acre than marshes and backwater wetlands.[46] But Mitsch and others would like to see a major restoration effort for these habitats, too. "We should have a policy of not plowing up to the edge of a stream," he explains. "We need this riparian zone to serve as a buffer, shade the stream, provide a bit of a relief valve for floodwaters."

As for the restored and created wetlands, they should be strategically located in two different parts of the landscape, he says. One would be right alongside rivers where wetlands such as the Olentangy or the Caernarvon marshes divert and cleanse part of the flow before returning it to the river or gulf. Such diversion wetlands would capture and cleanse the spring flood pulses that usually carry large amounts of newly applied fertilizers swept off the land by rains. The second location for new wetlands should be right on the farms themselves, intercepting nitrates as they're coming off the fields or out of field drainage systems. "The advantage of these is the nitrate concentrations are much higher the closer you get to the corn plant," Mitsch points out. "Therefore these wetlands can be more effective at nitrate removal, but on a smaller scale." A much larger wetland area is required to remove the same amount of nitrate once it is diluted in the river.

Even if wetland restoration and improvements in fertilizer management are implemented on the scale Mitsch and his colleagues envision, and the amount of nitrogen entering the northern Gulf of Mexico begins to drop, there could be a lag time of years before the dead zone shrinks.[47] Unfortunately, too, the level of nitrogen reduction required is a moving target. Population and food production in the basin are expected to continue growing, generating higher levels of waste.[48] Human-driven global warming may also aggravate coastal hypoxia, bringing more rainfall and thus greater nitrogen-laden runoff as well as hotter temperatures that intensify warming-induced stratification. One research team has calculated, for instance, that oxygen levels in northern gulf waters could drop by 30–60 percent if temperatures were to climb 7° F and Mississippi River water flowing into the gulf were to increase by 20 percent, a prospect that would greatly increase the extent of the dead zone.[49]

All of these trends emphasize the urgency of preserving and expanding the wetland habitat of the microscopic sediment creatures that help to remedy the ills created by our overfertilized civilization. And it confirms the folly of ignoring the power of vast numbers of little things underground.

VII

Fungi and
the Fate
of Forests

The lush coastal rain forests of the temperate climate zones rank among the rarest and richest forests on earth. Although remnants survive along the west coasts of Norway, Chile, Tasmania, and New Zealand's South Island, more than half of the world's remaining temperate rain forests stretch along the Pacific coast of North America from southeast Alaska down through British Columbia, Washington, and Oregon to northern California.[1] Cathedral groves of centuries-old Douglas firs still soar 15–30 stories high in places, creating damp twilight worlds where layers of cedar and hemlock, mosses and lichens, shrubs and spongy rotting logs succor mushrooms, salamanders, squirrels, birds, and bears.

Foresters who focused solely on timber production once pronounced such ancient conifer forests "over-mature," in need of cutting and rejuvenation. By the time ecologists began to challenge that view and reveal the vitality and complexity of old-growth forests, it was becoming apparent that these giants were being cut faster than anyone had realized. Nearly 90 percent of the temperate rain forests had been felled by 1990, reflecting a deforestation rate far greater than that in the better publicized and more extensive rain forests of the tropics.[2] A variety of economic, social, and scientific factors were converging, however, to bring a halt to rampant logging of Pacific Northwest old growth, from the needs of the endangered northern spotted owl to public protests and international boycotts of British Columbia timber.[3] Many public agency and corporate timber managers have now committed to reducing the cut and embracing a kinder, gentler ecosystem-based approach to logging called "new forestry."[4] This new approach recognizes, among other things, the need to protect forest soil communities, especially the vast underground web of mycorrhizal fungi that serves as an indispensable lifeline between forests past and future.

To the uninitiated, site C1500 on the east side of Vancouver Island looks no kinder or gentler than thousands of other clearcuts pocking the region. C1500 lies only 30 miles northwest of downtown Victoria, the capital of British Columbia, yet it takes nearly 2 hours for forestry technician Bob Ferris to negotiate a roundabout maze of rugged logging roads and pull his truck to a halt at this site in the Koksilah River valley. I climb out of the truck with Canadian Forest Service soil ecology research scientist Tony Trofymow and biologist Renata Outerbridge into a cold late October wind at the edge of a cutover hillside. This 160-acre site—about 1/4-mile square—is one of a half dozen on southern Vancouver Island that the Canadian Forest Service researchers have been studying to see how mycorrhizal fungi are faring under a partial-harvesting practice called "variable retention." This site had been blanketed by old-growth Douglas firs until 1999 when it was harvested by the timber company that owned it, MacMillan-Bloedel Ltd. That was the year after the company had pledged to phase out clearcutting in favor of variable retention. That commitment was later taken up by Weyerhaeuser's Coastal British Columbia Group, which took over MacMillan-Bloedel in late 1999 to become Canada's largest timber company. As we walk to the edge of the clearing, Trofymow begins to point out why, aesthetics aside, what we're looking at is not a classic clearcut.

One clue is a remnant patch of old growth rising from a rocky outcrop just upslope from the truck. The key to the new forestry, it turns out, is not just what's been taken away but also what remains. In variable retention, loggers cut at a range of intensities across a forested landscape, sparing patches or individual trees. On this site, Trofymow tells me, the crews left behind blocks of trees amounting to 15 percent (24 acres) of the original forest. Green trees are only the beginning, however. In the jargon of new forestry, what should remain amid the forest patches is a messy clearcut full of complex "biological legacies" from the old forest to help "lifeboat" the creatures, habitats, and processes that will enhance the recovery of tomorrow's forest. Legacies aboveground include not only live trees but also standing dead ones, downed logs, limbs, and shrubs. Equally vital are the legacies underground: roots, seeds, complex soil communities, and stocks of nutrients and organic matter.[5] I'm reminded of the "legacy carbon" that helps sustain life in the soils of the Antarctic Dry Valleys between wet seasons.

The legacies that foresters call "structure" are hard to miss as we walk upslope toward the patch of ancient trees above us, clambering over downed logs and other woody debris and pushing through low, dense salal shrubs, wild rose, and the serrated, pants-grabbing leaves of Oregon grape. The glossy oval leaves of salal are popular in flower arrangements, and Vancouver Island has become a prime area for pickers who sell salal branches to the florist trade. In the past, logging crews would have sprayed weed-killer on the salal and other shrubs and wildflowers to reduce competition with the replanted fir seedlings that dot the clearing below. Similarly, loggers once would have removed the slash or woody debris by burning the entire site. That practice has been abandoned here as much because of smoke and air-quality regulations as the need to retain downed wood and avoid scorching the thin forest floors on many of these slopes, Trofymow says.

Continuing upward, we enter a remnant stand dominated by tall, straight Douglas firs, their lowest branches well beyond our reach, and a few stately western hemlocks topped by droopy spires. The understory is sparse here, the rocky ground covered thinly with soil and expanses of reindeer moss. Despite its name, reindeer moss is actually lichen, Outerbridge points out, a symbiotic pairing of fungi and algae.

The trunks of the ancient firs around us are festooned with many other types of lichens as well as true mosses and liverworts, all denizens of the cool, damp forest interior. Will they survive here and help to recolonize the future forest despite the drying winds and sunlight that now intrude at the cut edges of this stand? New forestry is still as much art as science, and questions about how to monitor success are what have brought us to this site.

Outerbridge brings me a stick with bird's nest fungus growing on it, one of the same fungi I saw in the Smokies. A saprotroph—an organism that feeds on dead and decaying matter—this fungus absorbs its nourishment from decaying wood. She pokes around in the duff and holds up a small, thin-stalked *Mycena* mushroom, the fruiting body of one of the most common wood and litter decay fungi in this forest. Literally thousands of fungal species help to run these temperate rain forests by breaking down and decomposing wood and other recalcitrant debris, influencing community structure by attacking trees, or forming mutually beneficial partnerships as lichens, mycorrhizae, or endophytes. Endophytes make their homes unseen inside plant tissues and sometimes help protect their hosts from insects or disease. The Douglas firs all around us, for instance, harbor tiny fungal endophytes within their needles that produce chemicals noxious or toxic to grazing insects. And fungi also play vital roles in forest food webs. Many small rodents such as squirrels and voles eagerly consume truffles and other underground fruiting bodies, later depositing feces loaded with fungal spores and unwittingly helping to spread mycorrhizal fungi.[6]

It's those buried truffles and large fleshy mushrooms—and more important, the underground cooperative network of mycorrhizae that produce them—that most interest Outerbridge and Trofymow, and Weyerhaeuser as well. These are the fungi I've come to Vancouver Island to learn about. Their aboveground manifestations are one of the most sought-after commodities in these forests. Even now, mushroom buyers are stationed at Cowichan Lake another 20 miles west of here to purchase chanterelles from commercial foragers. These yellow-gold delicacies fetch a higher price per pound on the international market than timber, salal, or other natural goods harvested from the coastal temperate rain forests.[7] Outerbridge, who recently earned her doctorate studying the mushroom communities under various types of tree plantations, has a plastic bag of chanterelles, along with a few similarly prized matsutakes or pine mushrooms, sitting in her home refrigerator, the rewards of a weekend foray to Cowichan Lake. Trofymow, who earned his stripes as a soil ecologist probing how various assemblages of soil creatures affect decomposition and other processes in grassland soils, began collecting forest mushrooms to eat and taking his children on weekend mushroom-collecting forays years before his research for Canadian Forestry drew his attention to mycorrhizae. But gathering mushrooms is now merely a savory side benefit to their interest in these fungi.

Chanterelles, matsutakes, morels, porcinis, boletes, and many less tasty mushrooms are the reproductive parts—sporocarps or fruiting bodies—of mycorrhizal fungi—specifically, ectomycorrhizal (EM) fungi. Forest ecologists and managers alike now recognize that without the support of a rich legacy of EM fungi, no new forest would grow on this cutover site.

EM fungi are one of the three most common groups of mycorrhizal fungi. Their hyphae form a sheath called a "mantle" around a rootlet and deploy a network of silklike threads into the spaces between root cells. Like other mycorrhizal fungi, they project other hyphae far out into the soil, gathering and sharing water and nutrients such as nitrogen and phosphorus with their hosts in exchange for a share of the sugars the plant makes through photosynthesis. EM fungi are the ones that partner with the roots of firs, pines, and other conifers, as well as many other trees. Second among the common groups of mycorrhizal fungi are the arbuscular mycorrhizal (AM) fungi, whose hyphae actually penetrate the root cells of their host plants and do not form mushrooms. This is the most ancient and widespread group of mycorrhizal symbionts, and their hosts include most grasses and wildflowers, many of our most valuable crop plants, most tropical trees, and even the western red cedars of the Pacific Northwest forests. Third are the ericoid mycorrhizae that form on the roots of many forest shrubs such

Chanterelles and other highly prized edible mushrooms are the visible manifestations of vast underground networks of ectomycorrhizal fungi that form sheaths around rootlets and use their microscopic hyphae to extend the nutrient-gathering reach of tree roots. These fungal partners are essential for the growth of Douglas firs.

as salal and rhododendrons. All told, 90 percent of the plants in the world form cooperative partnerships with one or more species of mycorrhizal fungi. Some plant species can take the marriage or leave it, but Douglas firs must have fungal partners to thrive.[8]

Mycorrhizal fungi are mostly touted for their role in helping plants obtain water and nutrients, but these beasts provide a much wider array of services that helps to shape both the plant community and the soil community. (I say "beasts" with some justification because genetic evidence shows fungi are closer kin to animals than plants.[9] Although they appear "rooted" and immobile like plants, fungi don't photosynthesize or make their own food as plants do. And they stiffen their cell walls with chitin, the same material from which arthropods such as lobsters and beetles form their shells [external skeletons].) Mycorrhizae can form a maze of underground links between plants, using this hyphal conduit to share carbon and nutrients among them—for instance, doling out sugars provided by Douglas firs to hemlocks and other species growing in the shade below. Hyphae also exude some of these sugars into the soil, enhancing soil structure by binding particles into aggregates. Microbes feed on the exuded materials and on the hyphae themselves. Some mycorrhizal hyphae secrete enzymes that help decompose organic matter and even "mine" rock for mineral nutrients. Others produce antibiotics that protect plants from pathogens.[10]

We scramble down to the edge of the remnant old-growth stand where the trunk of a large Douglas fir has been ringed with blue tape and spray painted with an orange "T2." This tree marks the start of one of several transects that Outerbridge and Trofymow have marked, running from the forest edge out into the cut. Every 15 feet along these transects out to 150 feet from the cut edge, they have planted fir seedlings as "bait" to capture EM fungi that persist or are dispersed in the soil at various distances from the edge of the remnant forest. We work our way through the shrubs and debris well out into the clearing. Ferris helps Outerbridge dig up a foot-high fir seedling and bag its roots in clear plastic. We'll have to wait until we get back to the lab in Victoria and examine these roots under a dissecting microscope to learn the abundance and diversity of mycorrhizae that have colonized them.

Trofymow points out that this site was an early effort at variable retention, and it probably would not pass muster with company planners today. Retained patches are supposed to reflect the characteristic structure of the former forest. Yet the trees on the rocky ground above where their transect begins are probably smaller than the ones that grew on the deeper soil of the open hillslope, and thus not representative of the preharvest site. More old-growth patches should have been left standing on the productive soil of the slope, he believes. Neither cutting pattern would be likely to please those who want to see untouched forests here. But for Weyerhaeuser and others who plan to continue cutting trees, the main concern is not how the site looks but whether the patterns and practices applied here will sustain a rich array of forest dwellers, ensuring both the preservation of forest biodiversity and the success of forest recovery. One way to gauge the effectiveness of their practices is to monitor whether cutover sites retain the range of habitats, tree types and ages, and standing dead and downed logs found in old-growth forests. Another is to monitor how key plants and animals—sentinel organisms or bioindicators—respond to various harvest practices. Mycorrhizal fungi are obvious candidates, both because of their value to tomorrow's trees and also as part of the inherent biological diversity of these forests. What's more, mycorrhizae don't fare well in classic clearcuts, especially those stripped of debris and sterilized by fire and herbicides, because these fungi rely on aboveground greenery to supply them with energy.

A day earlier, I had gotten a glimpse of how the mycorrhizae are faring in this and the other Variable Retention Ectomycorrhizae Study (VRES) sites when I joined Trofymow, Outerbridge, and 20 other scientists in Nanaimo, a town 70 miles up the east coast of Vancouver Island from Victoria, for a meeting of Weyerhaeuser's Adaptive Management Working Group.

New forestry owes a great deal to the eruption of Mount St. Helens in 1980. The cataclysm scorched and leveled 230 square miles of forest in southwest Washington, leaving a visual wasteland of charred logs, volcanic rock, and ash covering the mountain slopes. Yet within 3 years, University of Washington ecologist Jerry Franklin and his colleagues were able to locate 90 percent of the plant species from the pre-blast communities surviving somewhere on those messy slopes.[11] Not just plants but also small burrowing animals, insects, soil denizens, and other creatures emerged from the destruction and set to work putting the system back on its feet. These findings, combined with several decades of research on the complex ecology of old-growth forests and the impacts of other natural disturbances such as fires and windstorms, convinced Franklin and others of the critical importance of biological legacies in ecosystem recovery.[12]

This realization, in turn, began to undercut a long-standing rationale that foresters had used to justify clearcutting: that it mimicked wildfires and other natural disasters. Unlike fires or even volcanic blasts, however, clearcuts simplify and homogenize landscapes, sweeping away most legacies, fragmenting the sites with roads, and replacing complex forests with the equivalent of tree farms. At a significant number of high-elevation clearcuts in the Pacific Northwest, in fact, foresters have not been able to get any trees to grow despite repeated plantings.[13]

It was Franklin who began to pull this information together in the late 1980s and publicize its implications under the catchy and controversial label of "new forestry."[14] He declared that timber production must no longer be the driving force in forestry. Instead, forested landscapes should be managed to mimic the complexity of natural forests and supply us not only with wood products but also biodiversity, wildlife habitat, productive fisheries, healthy watersheds, and not least, future forests.

But that is a philosophy, not a prescription. The devil, as usual, is in the details: What fraction of the trees should the chainsaws spare, how far apart, in what pattern, and in what habitats on any given landscape? Weyerhaeuser's Coast Forest Strategy, for instance, now zones its forests into three categories—timber, habitat, and old growth—depending on the values that will be emphasized.[15] The zoning determines the intensity of timber cutting allowed on a specific tract, as well as the proportion of the area available for harvest. Virtually all of Weyerhaeuser's privately owned land, such as C1500, is zoned as timberlands. Most of the land zoned for habitat or old growth is "crown" land leased long term from the provincial government. Besides its own decision to manage for nontimber values, the company must abide by a provincial forestry practices code that calls for special protections for streams, wetlands, some old growth, and other sensitive habitats. The result is that at the landscape level, a minimum of 20 percent of the forest in the timber zone, 30 percent in the habitat zone, and 66 percent in the old-growth zone will be retained.

Because new forestry is literally new, and because it can take 60 years or more to regenerate a mature forest—and at least 250 years to get old growth—no one applying these novel practices today will live to see whether they succeed. Thus, Franklin recommends treating each forest management practice as "a working hypothesis whose outcome is not entirely predictable."[16] That means employing what ecologists call "adaptive management"—a formal system for learning from the consequences of your actions. Learning requires monitoring consequences, and in this case that means keeping an eye on how sensitive groups of plants and animals are faring in various zones.

During the drive to Nanaimo, Trofymow had described one fundamental working hypothesis of new forestry that bears on the fate of life underground.

"One of the hypotheses used in planning cuts is that you shouldn't leave more than two tree lengths between retained trees or patches," he explained. "That's based on some research involving seed rain and dispersal of tree seeds. But are two tree lengths appropriate for all the other ecological values you're trying to sustain? What if you need no more than one tree length between patches for lichens? Or less than one length for some important soil organism or process? That's what this adaptive management effort and all the research is about. And if monitoring shows important functions and biodiversity aren't being sustained at two tree lengths, the idea is that they'll modify their practices and see if that helps."

But, I wonder aloud, will a timber company really want to change its harvest practices if snails or fungi or even birds don't seem happy?

"That's the $64 million question," Trofymow acknowledged.

One of Weyerhaeuser's stated goals has been to obtain third-party certification for its sustainable forestry practices in coastal British Columbia.[17] That seal of approval, in turn, should help it reap any marketplace rewards that come with being a "green" timber company. So in 1999, the company began submitting its plans and practices to annual reviews by an independent panel that has included Franklin and other prominent ecologists as well as representatives of major environmental groups. And Weyerhaeuser began developing what Franklin and others viewed as an innovative adaptive management effort.

The picturesque harbor town of Nanaimo got its start shipping out coal and logs from nearby mines and mills. Although it draws a growing parade of tourists, it still serves as headquarters for Weyerhaeuser's coastal timberlands, forest supply companies, and regional offices for provincial forestry and environment ministries. Many of the scientists attending the adaptive management workshop had spent season after season in the forests and cutover lands of Vancouver Island, listening for songbirds such as Townsend's warblers and golden-crowned kinglets, waylaying red-legged frogs and long-toed salamanders, searching damp places for tightcoil snails and jumping slugs, digging pitfall traps to census predatory carabid or ground beetles, mapping lichens and greenery, and examining tens of thousands of root tips for mycorrhizae. In the darkened room in Nanaimo, however, their findings were projected on screen as data points on charts and graphs, all configured to try to address tricky questions: What does 15 percent or 40 percent retention in patches "look like" to a snail, a bird, a beetle, or a fungus? Do any of them "care"? What does their presence or absence, abundance or diversity tell us about specific timber cutting practices? And what might we expect from them as the new forest grows and matures?[18]

Glen Dunsworth, then overseeing the adaptive management program for Weyerhaeuser, explained that the goal is to gather enough information on indicator species or groups to develop "species response curves"—that is, a graph that plots each creature's abundance or diversity or some other attribute against tree retention levels. "We want to know, is higher retention always better for a species, or do the returns level out at some point, so that, say, 25 percent doesn't do much more than 20 percent for the abundance of a certain species?" he told me during a lunch break. "And how do species attributes or abundances change as the matrix [the cutover site] ages?"

That afternoon, Outerbridge told the group that she had dug up 264 little fir trees along the VRES experimental transects since 2000. Back in the lab, she had clipped off a measured number of root tips from each fir—more than 40,000 root tips altogether—and examined them for the assorted colors and shapes assumed by ectomycorrhizae. Across all six VRES sites, Outerbridge had found 51 types of EM fungi. The seedlings planted in former old-growth sites such as C1500 had higher levels of root colonization and hosted a greater diversity of EM types than those planted in second-growth forest sites. Perhaps more important, the closer the seedlings were planted to the edge of a remnant forest patch, the greater the diversity of EM fungi on their roots and the higher the percentage of their roots that were colonized. That is what ecologists call a strong "edge effect."[19]

But the edge the fungi responded to turned out to be farther out into the clearing than the one I see as we pack up the freshly dug seedling at C1500 the day after the Nanaimo meeting. Outerbridge had found that the greatest declines in colonization and diversity actually occurred 50–150 feet out from the visible forest edge.

The key, Trofymow explains as we return to the truck, is that the roots of the old trees are extending out underground at least 16 feet or more into the clearing. It is another of the many unseen legacies that web this site.

"Those old trees we see at the edge of the stand have a root system that would be extending out, and then you have young trees whose roots would come in contact and become colonized with mycorrhizae through root-to-root contact," he explains. "And the roots of the young trees would then extend on and colonize other trees and so on in stepping-stone fashion," forming a vast nurturing network underground.

Some indication of what Trofymow and Outerbridge are likely to find in their research on how EM fungi fare at various levels of tree retention comes from a large-scale study in Washington and Oregon—the Demonstration of Ecosystem Management Options or DEMO project set up by the U.S. Forest Service to test new forestry concepts.[20] Oregon State University ecologist Daniel Luoma and his colleagues found that both EM fungal diversity and mushroom production improved dramatically from a clearcut to 15 percent tree retention to 40 percent retention. They were surprised to find in the 15 percent retention sites that mushrooms disappeared from under the retained green trees as well as from the clearing.

"The mycorrhizae were still there, totally intact, in the uncut patches," Luoma told me. "So the organism did not go away, but it stopped reproducing in the patches that were left behind." He speculates this was the result of a sudden change in wind, temperature, or moisture at the patch edges.[21] How long does it take for that shock to fade away? Studies like those at VRES and DEMO as yet provide only snapshots taken immediately after a harvest. "The dynamics of the recovery process are really the critical issue," Luoma said.

Trofymow also believes that the critical information is how the fungal community will recover over time: "We saw strong edge effects, but how long do they last? How long before the mycorrhizae colonize these clearings?" One way to get a glimpse of the answer without waiting for decades is to compare how fungi and other creatures are faring in forest sites of different ages. That's what Trofymow and his colleagues in the Canadian Forest Service had in mind in 1992 when they established a project called the Coastal Forest Chronosequence on Vancouver Island.

Years before timber companies and government agencies began to embrace new forestry techniques, ecologists working in the coastal rain forests of British Columbia had begun to wonder what else was being lost along with the old-growth forests and how long it might take for biodiversity and ecological processes to recover—if ever—in second-growth forests. By 1989, 40 percent of the land on Vancouver Island—including more than half of the mature old growth—had been clearcut and converted to "managed forests" of even-aged firs.[22] Within another decade, nearly one-fourth of the virgin forests along the entire British Columbia coast had been logged, creating 7.4 million acres (3 million hectares) of second-growth forest. Trofymow and other researchers have dubbed it, somewhat wryly, "our 3-million-hectare experiment on the coast"—an unfortunate test of what happens when an entire stage of forest development is eliminated.[23] The chronosequence project was set up to learn from that impromptu experiment.

Researchers picked eight sites on Vancouver Island for intensive study. Within an area no larger than 2 square miles, each site contains forest stands of four distinct ages: regenerating stands that were then 3–8 years old, immature stands 25–45 years old, mature stands 65–85 years old, and old-growth stands greater than 200 years old. For 5 years, teams of researchers haunted the sites, collecting detailed information on earthworms, carabid beetles, ground-dwelling spiders, EM fungi and mushrooms, nematodes, springtails, salamanders, lichens, and plants, as well as soil carbon and nutrient stocks. Their work would later help guide the design of Weyerhaeuser's adaptive management program.

The upshot of all this research is that even after 80-plus years, differences in biological richness persist between mature second-growth stands and old growth. A specialist looking at lichens on the trees and springtails and nematodes in the soil, for instance, will see a difference. And it is not certain whether the maturing stands, left uncut, will ever match all the characteristics of the old-growth stands nearby.[24]

For EM fungi, however, the result was different. Trofymow and then graduate student Doug Goodman found that the abundance and diversity of EM fungi in the mature and old-growth forests were quite similar, meaning that sometime after the first half-century or so, the EM fungi in the second-growth forests appear to have recovered.[25]

There are potential lessons for new forestry in that result: the mature stands used in the chronosequence EM fungi study were created by fire and early logging methods that inadvertently left some individual trees. Thus, the sites retained some of the legacies of the old-growth forests that had been cut and burned—hoary "veteran" firs still standing and large stumps and logs littering the floor, as well as rich stocks of organic matter and nutrients in the soil. More importantly, these disturbed forests sit alongside surviving stands of old growth from which EM fungi could disperse.[26]

Like C1500, the chronosequence site where much of the fungal work took place lies in the Koksilah Valley. After a half-hour on winding dirt roads Ferris pulls the truck to a stop at another dead end. We're still on timber company land, but there are no clearings in sight. To our right is a lush looking fir forest that Trofymow tells us was largely destroyed by wildfire about 90 years ago. On our left is old growth. Both are part of the Koksilah chronosequence.

Trofymow climbs out of the truck and charges into the dense understory of the mature forest, eager to locate the stakes and flagging that demarcate the now overgrown study plots. Outerbridge and I, both seeing the place for the first time and delighted by the contrast with C1500, follow more slowly into the green twilight. The wind sways the canopy far above us, but we cannot feel any stirring. Thick salal and woody deadfall cover the damp, spongy ground.

"Oh wow, look at him," I say, spotting a red mushroom.

"*Russula*," Outerbridge says.

I ask her whether it is an EM fungi.

"Yes, the whole genus is ectomycorrhizal. There are lots of species in the genus and some of them are good edibles. This one, no. It's not poisonous, but not good."

On a bank above a small tumbling creek we spot a whole cluster of little mushrooms. "This one looks like *Hebeloma crustuliniforme*, the poison pie," she says, crouching down for a closer look.

Mycorrhizal? I ask again.

"Yes, it is," she says, working her way down the soggy slope.

"There's reason to expect about 30 percent of mushrooms are ectomycorrhizal," Outerbridge explains. "A lot of it is circumstantial evidence based on where mushrooms are always found. But then we don't know, is it always under that tree because it's ectomycorrhizal on the roots or because it likes to eat that kind of litter?"

It's exceedingly difficult to dig around a mushroom or a buried truffle and track the hyphal threads back to their origin to see whether the parent fungus makes its living decomposing wood or eating sugars exuded by plant roots.

Traditionally, the taxonomy of fungi has been based primarily on the mushrooms or other reproductive structures. However, many EM fungi have not been linked to their corresponding mushrooms, so another taxonomy has been developed that uses the characteristic anatomical features and DNA profiles of the EM fungal mantle—the sheath formed around a rootlet—on a specific host. It's a bit like classifying fruit trees without linking them to the apples, pears, or other fruit they bear. Trofymow and others are trying to rectify that through an online database aimed at linking up DNA profiles, names, and descriptions of mushrooms with DNA profiles, names, and descriptions of the same fungal species forming ectomycorrhizae.[27]

At the creek edge, on the ragged end of a mossy, rotting log we find a clump of nondescript brown mushrooms. Outerbridge says they are one of several species of *Armillaria*, the honey mushroom, one of which causes shoestring root rot disease. This fungus attacks tree roots and sends ropelike bundles of hyphae known as rhizomorphs stretching across vast acreages in its quest for nutrients and new tree roots to parasitize. (These are the same "humongous fungus" mentioned in chapter 1 that may qualify as the largest and oldest organisms on earth.)

We cross the creek one at a time on rocks slick with green fuzz and walk up the far slope. Trofymow finds the center of one of the chronosequence plots, marked by a piece of rebar amid the shrubs. "Okay, this is Goodman's rebar here," he says, pushing the thick shrubs aside. "We didn't find any differences in terms of diversity of ectomycorrhizal fungi in this particular stand compared with the old-growth stand across the road," he recaps. He calls our attention to the tree trunks, all crusted gray with lichens. "This is where we found that the abundance and diversity of arboreal lichens increased from this site to the old-growth site across the road." For whatever reason, lichens either have not taken or cannot take full advantage of their proximity to a colonization source.

As we start down toward the creek again, Outerbridge finds *Xylaria*, a wood-decay fungus known as dead man's fingers. The name is apt. I see what look like puffy black fingers poking up from the duff. Before we reach the road she offers me a mushroom that smells like bleach, another that reeks sharply of garlic, and an edible *Russala* that smells faintly of herring.

When we enter the old-growth stand across the road, the sun is breaking through the clouds. I notice immediately that this forest is much more open, with a sparser but taller canopy of giant Douglas firs and tall hemlocks. When I remark on it, Trofymow consults his chart of stand characteristics. Everything on this site, it seems, has been quantified. Trofymow knows exactly how old this forest was when the chronosequence project began—288 years according to the cores he and his colleagues extracted from the biggest trees. Mean height here is 69 feet, and maximum is 118 feet, compared to a mean of 42 feet and a maximum of 90 feet across the road. Mean density of trees here is 193 per acre compared to 1,357 per acre across the road. A more open forest indeed.

When we arrive back at the Canadian Forest Service Pacific Forestry Centre in Victoria in late afternoon, Outerbridge gives me my first close-up look at the EM fungi I've been hearing about for 2 days. On a laboratory counter she pulls the fir seedling dug up at site C1500 from its plastic bag, plops the ball of roots and dirt into a white plastic tray, and begins gently separating the roots from the soil.

"See these white threads here?" she asks, pointing to what look like masses of spider webs lacing the dark soil. "Those are hyphae." They break up quickly as she works.

Outerbridge cuts off the ends of a dozen roots, rinses them in a sieve, and puts them in a tray of water under a binocular microscope. Peering into the eyepiece, I spot what look like bits of fuzzy white yarn around the orangey root tips.

"That's *Pseudotsugaerhiza baculifera*. It's sort of a tentative name," she says. "It just means 'Douglas fir mycorrhizae' and indicates a descriptive feature 'bearing small twigs,' probably referring to the needlelike crystals on its hyphae."

At least that creature has a name of sorts. Most of the EM fungi found on her seedlings remain unidentified. She creates descriptions and photographs and refers to them as: "felty translucent white," "chocolate-brown metallic," "pink," "purple blue," "Tomentella-like," or "reddish brown, pubescent, monopodial pinnate to pyramidal in one plane."

What looks like cottony yarn is a fungal mantle; the fuzz is a fan of hyphae.

"Even the finest root hairs of most plants are fatter than most fungal hyphae, and they are fewer in numbers and shorter, so overall they have less surface area," Outerbridge points out. "The idea is to increase the surface area of the root tip for better nutrient absorption." The hyphae can also explore and exploit the tiniest nooks and crannies of the soil where root hairs cannot penetrate.

She projects the microscope view onto a computer screen so that we can both see the next one, a black, tuberous-looking bump that she identifies as *Cenococcum geophilum*. This is one of the two most common EM fungi she has found on the young firs. "In this sample I'll probably find 40 percent of the root tips occupied by *Cenococcum*," she says.

"I'm surprised I'm not seeing *Rhizopogon* in this sample because it's usually the most common." She keeps moving the tray slowly. "Oh, here it is."

The brown blobs? I ask.

"Yes, it's a blob because it's a tubercle type. Basically this whole fungal mantle you see is coating several root tips like a mitten over fingers." She cuts into the tubercles with forceps and we can see the root tips bundled inside.

On a nearby root she sees another EM type, a fuzzy yellow-brown ball that looks like a pinch of fiberglass insulation.

What we are seeing is a tiny sampling of EM fungal diversity. Researchers have identified 2,000 species of EM fungi just from Douglas firs.[28] A tree might partner with 10 or more EM types at any given time in its life.[29] Some fungi may colonize only a specific tree species, or trees at certain stages of life. For example, Outerbridge has found no chanterelle fungi on any of her 40,000 seedling root tips.

This raises the question of just how many mycorrhizal types a healthy forest needs. What is the threshold below which the integrity of the soil community and thus the health and productivity of the future forest is compromised?

I asked this of David Perry, an emeritus professor at Oregon State and a pioneer of old-growth forest ecology, who has taken part in the annual evaluations of Weyerhaeuser's Coast Forest Strategy. He laughed, then sighed.

"When you compare the soil of any retained forest with a clearcut, you always find differences," he said finally. "We just don't have enough research and experience yet to sort out what that means in the long term for timber production or even for biological diversity. It's pretty clear you can lose too much, but what is too much varies from site to site. We've got a lot of detective work to do to sort this out."

Perry's own work supplies clear cautionary tales from a number of harsh sites where some threshold was crossed and the soil lost its ability to grow a forest. At a fair number of droughty, high-elevation sites from Oregon to Montana, foresters have failed repeatedly in their efforts to get new trees established on clearcuts. The most intensively studied is Cedar Camp, a site in southwest Oregon where Perry and his students spent years trying to determine what triggered degradation so severe that the soil turned to beach sand.[30] His first assumption was that the EM fungi had been lost, but it turned out to be more complicated.

"The ectomycorrhizal fungi were actually still there," he recounted. "But they had lost the ability to form mycorrhizae with the tree seedlings. We still don't fully understand that." Perry believes the trigger was probably the herbicide spraying of young shrubs and hardwoods that sprouted in the clearing. Shrubs and hardwoods not only serve as legacy plants that "lifeboat" some of the important mycor-rhizal fungi but also seem to suppress a common soil microbe, the actinomycete *Streptomyces*. *Streptomyces*, from which we got the antibiotic streptomycin, chemically suppresses many other microbes and plants, and its proliferation probably reduced root growth and formation of mycorrhizae on replanted fir seedlings. New plantings died year after year as seedlings apparently failed to gather enough nutrients and water to survive their first hot, dry summer. Without trees or shrubs to pump carbon-rich sugars into the soil, the underground food web that depends on that aboveground subsidy began to unravel. The EM fungi hunkered down and went dormant. Deprived of sugars, roots, and hyphae, soil structure deteriorated. The critters most devastated, apparently, were the grazers such as mites that devour microbes and release nitrogen that fertilizes plant growth.

Mike Amaranthus, at that time a Ph.D. student of Perry's, was finally able to break the cycle and get trees growing by planting seedlings in soil taken from an established forest—most likely, Perry believes, because the transplanted soil reintroduced grazers, and their munching of microbes jump-started the nutrient cycle again, allowing the seedlings to grow enough roots to finally take on mycorrhizal partners.[31]

The full story is much more intricate, as are most tales from life underground. But the clear caution to foresters and land managers is to avoid breaking the links and legacies that allow plants and soils to nurture one another. Ensuring the survival of a diverse soil food web is critical to the recovery of the forest.

It remains to be seen whether Weyerhaeuser and other timber companies and public agencies will continue to monitor the impacts of their new forestry practices and—more important—refine these practices based on the results. The outcome has meaning far beyond the future productivity of our forests. The integrity of forest soils affects a wide range of other aboveground processes, from release of carbon to the atmosphere to erosion, nutrient losses, and water quality, in the immediate watershed and also far downstream.[32]

The outcome of changes in forestry also has meaning for the biological diversity the forests harbor. Forest issues in British Columbia have attracted keen interest worldwide precisely because the rich plant and animal life of the province remains largely intact.[33] Despite a century of intensive timber cutting, wolves, black bears, and cougars still inhabit Vancouver Island. Outerbridge has heard the howl of a wolf as she worked at C1500 in the Koksilah Valley. Will these large and charismatic creatures continue to find suitable forests to live in as second-growth forests on the eastern side of the island and old-growth forests on the west continue to be cut? Most of what little old-growth Douglas fir–dominated forest remains on eastern Vancouver Island exists on privately owned land that is unlikely to be turned into parks or preserves. Because of that, conserving the biodiversity and other values embodied in natural old-growth coastal rain forests in this region depends largely on the success of today's experiments in kinder, gentler forest management.[34] Today's timber harvesting practices must be designed to protect life and processes both above and below the ground while the forest recovers, and only by monitoring the fate of key forest species and altering forest practices accordingly can we gauge whether we are succeeding. If we continue to pay attention to and learn from vital but long-overlooked creatures such as mycorrhizal fungi—and act on what they're telling us—perhaps tomorrow's forests can be as complex, lively, and awe-inspiring as the forests we are cutting today.

VIII 🪲

Grazers, Grass, and Microbes

Yellowstone National Park has been called "America's Serengeti" for a unique wildlife spectacle that rivals the annual migration of vast herds of zebra and wildebeest across the East African savanna. Each winter, elk and bison congregate by the thousands in the northern third of Yellowstone—the relatively low-lying, sagebrush-dotted grasslands that stretch along the Lamar, Gardner, and Yellowstone rivers and extend across the northern park boundary into Montana. There the animals paw away the snow to expose dried grasses that sustain them for up to 7 months. Around April, when the first flush of green grass emerges on the northern range, the animals graze it heavily and then begin migrating south, following the retreating snow and the advancing wave of greenery upslope.

Within decades after Yellowstone was declared the world's first national park in 1872, some managers and scientists became convinced that these wintering herds were too numerous for the health of the northern range. Perception that the winter range was overgrazed intensified with the drought of the 1930s and has persisted in some quarters ever since. For almost a century, managers aggressively controlled elk numbers, as well as fire and predators, eliminating gray wolves by 1926 and reducing elk numbers to a low of 4,000 by 1962. Soon thereafter, national television images of park rangers shooting elk sparked a public outcry. In 1968, yielding to public pressure as well as shifts in ecological thinking, Yellowstone adopted a largely hands-off policy of "natural regulation," relying on natural processes inside and hunters outside the park to limit elk numbers. Two decades later, the elk herds reached highs of 19,000 and the debate continues among range managers, soil scientists, wildlife biologists, and ecologists about whether elk and bison are degrading the northern range.[1]

The controversy is part of a much larger conflict over evolving ideas of how to manage grazing lands sustainably, for livestock or for wildlife. The outcome is vital because grassland ecosystems cover about one-fourth of the earth's land surface. Barely two centuries ago, most of these grasslands still sustained large free-ranging herds of wild grazers, whether zebras or elk, saiga antelope on the Eurasian steppe or ecologically equivalent but hoofless kangaroos on the Australian savannas. According to the fossil record, wild grazers such as these have coexisted stably with grassland ecosystems for some 70 million years. Now, however, most of the earth's grasslands have been plowed up for crops or converted to range or pastureland.[2] Cattle and sheep have degraded rangelands in many parts of the world in a matter of decades, stripping away plant cover, compacting the soil with their hooves, and leaving bare ground open to wind and water erosion. Today, Yellowstone and Serengeti are among the few surviving semi-natural grasslands where we can learn how grass and grazing animals coexist—lessons that link to life underground and that may help us better manage pastures and rangelands for the future.

The first such lessons began to emerge from the Serengeti in the 1970s, and they were startling. Serengeti National Park in Tanzania and the adjoining Masai Mara Game Reserve in Kenya form a tropical grassland three times the size of the temperate grasslands of Yellowstone and host 50 times as many ungulates (hoofed grazers). Yet the Serengeti's grasses are not passive victims of the grazers. In turn, the zebras and wildebeest that migrate by the millions across these plains do not rely passively on the rain and grasses to determine the amount of forage available to them. Instead, as ecologist Sam Mc-Naughton of Syracuse University showed, these animals actually stimulate grass production. On average, the grazed grasses produce twice as much greenery as ungrazed grasses, and that greenery is younger, denser, and more nutritious in grazed areas.[3] Thus, the animals greatly increase the carrying capacity of their own habitat.

"It was some of the most exciting work I'd ever encountered in ecology," Doug Frank says, telling me how he had come across Mc-Naughton's findings in the 1980s. It is a June morning and we are chatting in front of the historic stone post office at Mammoth, the headquarters of Yellowstone.

"He was discovering all these really cool things about how plants and large herbivores in East Africa were highly adapted to one another, and I decided I wanted to see whether these same kinds of feedbacks are occurring in temperate systems like Yellowstone. It's really the only place left where we can get a glimpse of the kinds of ecological processes that occurred throughout the earth's temperate grasslands in prehistory."

At the time, Frank had a master's degree in plant ecology and was studying the effects of mountain goat grazing on alpine plants in Olympic National Park. McNaughton agreed to take him on as a Ph.D. candidate, and Frank spent the next 4 years working in the park each spring and summer as a graduate student. He has been coming back ever since. "This work continues to be very exciting," says Frank. Each season when he arrives in the park, "it's like coming home."

Today he is letting me tag along to one of the northern range field sites he has been using for the past 15 years. As we drive up the winding road past the crowds of tourists at Mammoth Hot Springs, Frank, now an associate professor at Syracuse alongside his mentor Mc-Naughton, recaps what he found in those first years.

The bottom line is that on Yellowstone's northern range, as on the Serengeti, grazing doubles the amount of plant-available nitrogen in the soil and stimulates more abundant and denser grass growth.[4] Grasses that have been chomped by elk or bison produced

45 percent more greenery on average during the growing season than ungrazed grasses, Frank found. Further, his findings came in a drought year when the elk population had reached a high of 19,000 and bison, too, were at record numbers.[5]

"Once you see this very strong positive feedback effect that the animals are having, you have to begin to ask 'okay, but how in the world is this happening?' It really makes no intuitive sense. And that's when we started looking at the soil."

Historically, grazing research has focused on the animals themselves —everything from population densities to bite size to food preferences —and on the plants they consume—the makeup and diversity of the plant community, plant defenses, productivity. Now a growing number of researchers have come to realize that the soil community plays a crucial role in choreographing the complex duet of grasses and grazers. Plants are intimately linked to the soil community through feedbacks involving energy (carbon) transfers, decomposition and nutrient cycling, symbiotic relationships with mycorrhizal fungi and nitrogen-fixing bacteria, and inhibition by soil pathogens and root-feeders, as we've seen throughout this book. It's now increasingly clear that grazing animals, too, from elk to grasshoppers, can influence the feedbacks between plants and the belowground community.[6] And humans alter this interaction by reducing or increasing populations of grazers or introducing nonnative grazers to the landscape.[7] Thus, in the early 1990s, Frank and a number of other ecologists found themselves pulled, unexpectedly, toward the opaque frontier of the soil.

After only a few minutes, Frank pulls over to the left side of the road and parks the car in a small turnout. It's a quixotic summer day in the Rockies, windy and pleasantly cool, with sudden rainsqualls alternating with what is now brilliant sunshine. Before us, a grassy slope dotted with sagebrush drops down to a small plateau bordered by a dark stand of Douglas fir trees. No animals graze on the plateau.

"This site is grazed intensively in the springtime," Frank points out. "As the snow melts and this area begins to green up, we see a lot of evidence of grazing." But it is June, and the animals have moved up into the mountains and high meadows.

The perennial grasses on the plateau before us still look green and plentiful, and I wonder aloud why the animals migrate so far and return only when this grass has dried to hay and been buried by snow? It's a question Frank and his colleagues have spent years answering.

Essentially, the animals choose the richest, most productive sites in this dramatically variable landscape to graze, he explains. By measuring the total green matter—net aboveground primary productivity—a patch of land produces, and comparing that with how much the animals consume, he found that the more productive a patch is, the more of its grass the animals eat. And they eat the actively growing young shoots as soon as the grass begins to sprout, not waiting for large masses of greenery to accumulate.

"In Yellowstone, the growth of grasslands comes in a pulse of about a month and a half, and the migratory grazers track this pulse, this green wave of young tissue, from low to higher elevations, moving from winter range to higher summer range," he says. "We can see the migration, but we wanted to know what kind of benefit the animals are deriving from it."

The answer he's pieced together over the years is that the animals get more food and better nutrition per bite and thus a better diet for less effort by migrating. It turns out that the nitrogen and mineral content of grass blades on this range peaks a few days after green-up begins in the spring. And the young grass is denser and more highly concentrated than taller, older grasses, allowing the animals to get more of this highly nutritious food per bite.[8] Efficient foraging is particularly important in spring when females are bearing and nursing calves.

On the Serengeti plain, migratory herds move seasonally across more than 100 miles of rolling savanna, following the rains and the "green wave" of plant growth that begins in Kenya's Masai Mara and sweeps southeast into Tanzania as the wet season progresses. Just as in Yellowstone, McNaughton has found that this pattern of movement allows young animals and pregnant and nursing females to graze forage with elevated levels of necessary minerals.[9]

At the richest sites in both Yellowstone and the Serengeti, the grazers may consume more than half of the grass produced each year. McNaughton, Frank, and their colleagues scoured the scientific literature to compare the amount of green matter produced with the amount consumed in a wide range of other ecosystems around the world. The exercise revealed that ecosystems fall into two very distinct categories. One group contains low-herbivory systems such as desert, tundra, temperate forest, tropical forest, and small remnant grasslands without large grazers. Here, on average, animals large and small consume only about 9 percent of the vegetation produced each year. The second group includes high-herbivory systems, mainly large grassland regions such as Serengeti and Yellowstone that still support abundant migratory herds, where consumption averages 55 percent.[10]

"How can an ecosystem that experiences this kind of intense, high, chronic herbivory be sustainable?" Frank asks. "Because this appears to be the case if you believe the fossil record." This is the overarching question that has brought him back to the park year after year.

As we begin walking downslope, I catch a glimpse of a wire-fenced enclosure below us, not readily visible to visitors from the road. Actually, scientists call it an "exclosure" because the fence is designed to keep elk, bison, and other large grazers out. It's a simple way of allowing researchers to examine grass production and other ecological processes in the same ecosystem with and without grazers. This 5-acre exclosure is one of eight scattered across the northern range from Mammoth east through the Lamar Valley. All were erected more than four decades ago, just a few years before the park stopped culling elk, and all serve as testimony to the long-running controversy over the impacts of grazing here.

What's most striking looking down from above is a contrast that rankles park critics. Dense thickets of 20-foot-tall willows cover much of the ground inside the fence. In the meadow beyond, where elk, bison, pronghorn antelope, and mule deer have been free to feed for 40-plus years, we see only what Frank describes as "little nubbins of willow" here and there.

At the foot of the slope we walk out across the meadow and approach the tall wire fence. The original wooden posts now sag outward in places, and newer metal fence posts have been placed at angles to buttress them. Nearby, a small stream trickles out of the exclosure. The ground is spongy and damp, green with Kentucky bluegrass, rushes, clumps of wild iris, and a yellow-flowering cinquefoil. Inside the fence, wild rose bushes and irises crowd the openings in the willow thickets. The grass inside appears no higher than out, but dense gray thatch and standing dead grasses mingle with the greenery inside. Frank points out that far less litter accumulates out here in the grazed meadow. The result is dramatically warmer soil temperatures and increased soil microbial activity during the growing season, and colder winter soil temperatures in the grazed areas.

As we walk along the fenceline, Frank explains that soggy meadows or swales like this one readily support willows. If grazers are around to browse the growing tips, however, "a willow will survive, but it's going to remain small." Indeed, across the northern range, grazers have visibly altered the pattern of trees and shrubs since the early 20th century. Willows are short and sparse, and all tree-sized aspen and cottonwoods got their start before the 1920s. Elk browsing has prevented "recruitment"—growth of seedlings or suckers into mature trees and shrubs. And browsing by elk, pronghorn, bison, and mule deer has reduced the extent and height of sagebrush at lower elevations on the winter range.[11] Those changes, along with questions about the state of the grasses and the soil, underpin the controversy about whether the herds are degrading the winter range.

Various experts still make dramatically different judgments about the landscape in which Frank and I are walking. To understand how some perceive ruin where others see vibrant health, it's necessary to delve briefly into the changing world of range science. In the 1940s, range scientists developed a system of rating the state of grazing lands that was based on then-current ideas in ecology about what a plant community on a given site should look like. Basically, the thinking was that a plant community on a site develops through a predictable succession of stages toward a single idealized climax or stable state unless it is disrupted, say by excessive grazing.[12] The heavier the grazing, the more plant community succession is held back. Range specialists classified the condition of rangeland such as this plateau based on the percentage of the vegetation that matched what was believed to be the natural climax state. If too many climax elements appeared to be missing—say willow and aspen in the case of Yellowstone—just reduce or eliminate the grazing pressure, the idea went, and the damage would be reversed and succession toward a predictable endpoint would be resumed.

Ecologists have since moved away from the notion that all plant communities develop inexorably toward a single stable state, and for the past several decades, range scientists have been struggling over changing standards for interpreting what they see and classifying and managing range conditions.[13] Nevertheless, with the elk population soaring in the 1980s, the debate over the state of Yellowstone's winter range grew loud enough that Congress in 1986 directed the National Park Service to "start a study on Yellowstone to see whether there is evidence of overgrazing [and] what should be done to avoid that."[14] Work by Frank and a bevy of other researchers has mushroomed since that time. In 1998, again at the request of Congress, the park service asked the independent National Research Council to assess the findings to date.

The research council's 2002 report confirmed that the "condition of the northern range is different today than when Europeans first arrived in the area," and larger numbers of elk and bison share responsibility for that along with human development and possibly shifts in climate. But the report concluded that no major component of the ecosystem was in danger of being lost: "[A]lthough we recognize that the current balance between ungulates and vegetation does not satisfy everyone—there are fewer aspen and willows than in some similar ecosystems elsewhere—the committee concludes that the northern range is not on the verge of crossing some ecological threshold beyond which conditions might be irreversible." And the park's "practice of intervening as little as possible is as likely to lead to the maintenance of the northern range ecosystem and its major components as any other practice."[15]

As for the grass: "The composition and productivity of grassland communities in the northern range show little change with increasing grazing intensity."[16]

Park critics were far from mollified, but in the winter of 1995–1996, a new twist was added to the ongoing saga. Gray wolves were reintroduced to the park after a 70-year absence and quickly flourished. Elk counts from the 1999–2000 to the 2002–2003 winters ranged from 11,700 to 14,500 animals. Now researchers are seeing signs that young willows, aspens, and cottonwoods are taking hold on the northern range, and not simply because wolves are helping to reduce elk numbers. A new wariness about wolves may also be driving elk to change their behavior since they can no longer linger fearlessly to browse along river bottoms and in open meadows where they make easy prey. Ironically, the too-many-elk voices are now being drowned out in the local media by hunters and outfitters who fear that wolves will eat so many elk that there will be none migrating out of the park in fall and winter to shoot.[17]

With wolves and elk in the public eye, new findings about how the soil community influences the interactions between elk and their food plants have gotten little attention.

Frank has come to the Mammoth exclosure today to arrange for some soil sampling, and he wants to avoid taking samples from sites near the fenceline that were fertilized with nitrogen during a previous project. He drops to his knees and begins crawling along the fenceline, poking his gloved hands into the grass under the wire to feel for metal pins that mark the boundaries of the old plots. I drop to my knees and grub in the grass, too, unsure what the pins might look like. It takes 10 minutes of crawling and searching before Frank spots the first crumpled and weathered copper identification tag attached to a pin. I get down inches from the ground to read the numbers etched on the tag.

"It looks like 'Study #GLL01668'," I report. "Then it's got a number like 1-0034."

"Could that be 0037?" he says, studying a plot diagram. Yes, I allow.

Based on that number and the diagram, we search for the other corner pins.

This site and a handful of others throughout the northern range have been sampled and probed extensively over the past two decades, but except for the corner pins, few tangible signs of the studies remain. Since the mid-1990s, Frank and his colleagues have been delving beneath this ground to find out just how, as the National Research Council noted, these grasses can be so little affected by being literally half eaten each spring.

One plus for the grazed grasses is the dung and urine the animals leave behind. Grass wastes processed through the guts of elk and bison are much more easily broken down by decomposer microbes than uneaten grass litter. Thus, nutrients in these waste patches are rapidly made available to fertilize new plant growth. But Frank and others believed that this effect was too spotty to explain the luxuriant boost in grass growth they were documenting across the grazed landscape. The grazers had to be triggering some other positive effect belowground.

"It's kind of like peeling the layers of an onion," Frank explains. "You figure out the exterior layer, like enhanced productivity with grazing, but that's being controlled by something below. So you try to peel that next layer away just to get at the nexus of what might be happening. It's all very exciting."

And that layer led you into the soil? I ask.

"Kicking and screaming," he says, laughing easily. "The problem is, that's really hardcore biogeochemistry. If what turns you on is being out in the field, then when you decide that you want to do soil process work you have to spend an awful long time in the laboratory with pot experiments and extracting soils with dilutions, all that stuff."

Fortunately for Frank, "an extremely talented lab junkie" named Bill Hamilton was working as a postdoctoral researcher in his lab when he decided to find out just what happens belowground at sites like this each spring. Since herds graze a site early in the growing season before moving on, the ground is still moist and spring conditions are favorable for the grasses to rebound and compensate for the lost greenery. The question is, how do they go about it?

Hamilton took samples of Kentucky bluegrass (*Poa pratensis*) and fresh soil from a grazed area like the one where we're crawling around. In a greenhouse at Syracuse, he grew the grass in pots filled partly with field soil so that their microbial communities would be similar to those under grazed grasses on this range. Then he clipped the grasses with scissors to simulate grazing. To track what happened next, Hamilton spiked the carbon dioxide the potted grasses were breathing with a carbon-13 (^{13}C) isotope.

You might imagine that the first act of a wounded plant would be to devote all the carbon it could capture through photosynthesis to building new greenery in order to capture more light, grow more stems and leaves, and so on until it had compensated for the lost growth. But as Hamilton tracked the fate of the ^{13}C with a mass spectrometer, he found instead that all the action went underground.

Within 24 hours, the clipped grasses were pulsing five times as much sugar and other carbon-based substances from their roots into the soil as the unclipped grasses. This sudden bounty quickly spurred a dramatic population boom among soil microbes in the rhizosphere adjacent the roots, and these microbes set to work decomposing organic matter and releasing nitrogen and other nutrients. Later work showed that the increased root exudation and microbial activity lasted about 3 days before starting to taper off. But that frenzy of decomposition and increased nutrient release was enough to launch the plants toward regrowth and recovery. After a week, the clipped plants enjoyed higher nitrogen uptake, higher nitrogen content in their leaves, and a 24 percent higher photosynthetic rate than unclipped plants.[18]

Feeding extra sugars to soil microbes must be quite an expensive investment for an injured plant, I comment.

"Yes, so it's a very short-term pulse, and that makes sense," Frank replies. "But grasses exude a lot of carbon anyway, up to about 20 percent of the carbon that they assimilate, so it must be important to them." The strategy of ramping up carbon exudation in response to grazing makes sense if the grass finds nitrogen in shorter supply than carbon.

"That seems kind of counterintuitive," he says. "But I think that's what's happening. Before the plants are grazed, their growth is limited by the amount of carbon they can assimilate through photosynthesis. So they're concentrating on producing stem material and trying to grow tall and reach the light, in competition with their neighbors. Once the herd goes through and the plants are grazed, they're all grazed to about the same low level and there's less competition for light. So carbon may not be limiting any longer. What become limiting are the soil resources. And I think that's why you see grasses after they're grazed allocating this carbon resource belowground."

Nitrogen, as gardeners know, is the primary fertilizer of plant growth, and it's a limiting resource in all grassland ecosystems.[19] Several studies have found the same postgrazing response in other members of the grass family, including corn plants and blue grama grasses attacked by munching grasshoppers,[20] and even grassland plants attacked by root-feeding nematodes.[21]

"It appears that this is a robust response that grasses have to being defoliated," Frank explains. "So the question was, why are grasses doing this? They're losing their photosynthetic machinery by being grazed. It would seem that carbon is not what they want to lose, so why are they exuding it?"

Other studies had found that grazing can increase the number and activity of microbes in the rooting zone.[22] By labeling the carbon and tracking the extra pulse into the soil, Hamilton was able to link it directly to both the microbial population boom and accelerated nutrient cycling.

The plants seem to get a second payoff on their investment when the microbial masses die or go dormant. "Adaptively, it makes sense for the plant to cultivate this large population through exudation,

Grasses grazed by bison or elk in Yellowstone National Park quickly ooze extra sugars from their roots, spurring activity throughout the soil food web, accelerating nutrient cycling, and enhancing release of nitrogen that fertilizes new grass growth. Dung and urine patches also speed decomposition and nitrogen release.

then cut off the supply. The gravy train is over, they die, and you can get another big pulse of nitrogen out of the decaying microbes."

Of course, a microbial community in the soil doesn't boom in isolation. Tiny predators such as nematodes and protozoa arrive to graze on the microbes, and other soil animals such as mites or spiders consume these tiny predators, and so on, so that the nitrogen in microbial carcasses may cycle through a complex underground food web before becoming available to fertilize plant growth. That happens when decomposer microbes finally break down the carcasses of these soil creatures and release the components into the soil in inorganic forms such as the nitrate that plants require. Although Frank and Hamilton did not investigate this cascade, an earlier study on the Yellowstone winter range found that the density of both bacterial-grazing and root-feeding nematodes was greater under grasses grazed by elk and bison than under ungrazed grasses.[23] And others have reported increased activity by microbe-consuming soil animals in the rhizosphere of grazed plants.[24] The consumption of microbes by nematodes, protozoa, or springtails actually increases the growth of both microbial populations and plants.[25] Indeed, underground consumers such as nematodes can speed up nitrogen cycling by 20–50 percent over systems with only microbes.[26] Thus, elk and nematodes may unwittingly interact to stimulate the growth of plants on whose productivity both ultimately depend.

"My big picture take is, plants are capable of manipulating the microbial activity in the rhizosphere to increase nutrient availability, and they do that in response to grazing," Frank sums up.

But the amount of carbon that grasses share with their microbes, routinely or in times of stress, is just one indication of how much effort grasses here expend on cultivating their ties with the soil. Frank reminds me that grasses always invest more effort in building roots than greenery.

"There's always more biomass belowground. Always. In grasslands, most of the allocation is belowground because water is a big limiting factor, so they have to allocate a lot of biomass to roots in order to get water," he explains. Nitrogen, too, as we've just seen. Yet until recently, most studies have shown that aboveground grazing reduces root production.

Frank and his colleagues recently discovered that, to the contrary, Yellowstone grazers stimulate productivity belowground as well as above. At this exclosure and others across the northern range, his research team sank clear tubes known as "minirhizotrons" into the soil both inside and outside the fences. Using a tiny video camera and a light source inside the tubes, they were able to zoom in on a field less than 1/10th of an inch square and photograph individual grass roots growing along the outside of the tubes. By photographing the same roots month after month and digitizing the photos back in the lab, they were able to calculate the growth or shrinkage of each root during the season.

"It's painfully time-consuming, but it's really the only accurate way to monitor belowground production," Frank says. "And to our great shock, we saw greater root production as well as aboveground production in the grazed plots. In fact, the rate of stimulation by grazing was seven times greater belowground than aboveground."[27]

When you put all these bits and pieces from 15-plus years of research together, it's possible to visualize a remarkable cascade of events taking place largely unseen on this range each spring as elk and bison crop the grass low and leave. The timing is just right: The soil is still moist from snowmelt, the sun is quickly warming the long-frozen ground, and the growing season is young. Structurally, the grass is primed for growth: Its growing tips hug the soil surface, safe from the teeth of grazers and ready to send out lateral shoots, or tillers. The taller, older greenery has been stripped away, allowing the new shoots with their higher photosynthetic rates greater access to light. The herds have tilled the earth with their hooves. Belowground, microbial populations are burgeoning as the creatures feast on highly digestible carbon: not only the extra bursts of sugars, carbohydrates, and proteins exuded by the roots of the injured grasses but also patches of dung and urine deposited across the surface by the animals. The microbes are eating and being eaten, fostering a complex underground food web that accelerates the nutrient cycle and frees up generous amounts of nitrogen. The grasses, in turn, soak up sunlight and nitrogen, build new roots, and recover their lost greenery.

After crawling around in some of this postgrazing greenery for a half hour, we've located only three copper-flagged pins and one unflagged marker. But Frank feels that's enough to guide future sampling and avoid the fertilized plots. Besides, the sky is darkening and the wind is picking up.

As we head back upslope to the car, I ask whether he's excited about the soil organisms that have begun to enter his work and his life.

"Oh yeah, I am, very much so," he says, laughing easily. "I've just arrived there, realizing the soil is an important key to how everything else in the system is operating."

Each finding, it seems, is taking him farther underground. After discovering how grazers and grasses can manipulate activity in the soil community, Frank had begun to wonder whether thousands of years of such interactions might have altered the character and identity of the soil community in grazed grasslands. That's the layer of the onion he's peeling back now.

We reach the car and drive back downhill to the restaurant at Mammoth. Soon after we're seated, the sky darkens and a curtain of rain sweeps across the landscape, settling in for a long downpour that drives gaggles of wet tourists inside to join us.

I ask him about the next layer of the onion, which began with an idea for an experiment that came to him while attending a workshop on plant-herbivore interactions in Sweden.

"I remember describing this pot experiment to some folks over beers, and they said 'ah that's never going to work,'" he recounts, laughing. "They thought it was kind of a harebrained notion, but I did it anyhow, and it turned out really interesting. We asked a simple question: Do grazers affect the composition of the soil microbial community?"

As soon as he returned to Syracuse, Frank arranged to have soil samples and grasses taken from inside and outside the Mammoth exclosure we have just left. The soil samples were chilled and shipped overnight along with the grasses to his lab in New York to ensure that the soil community remained intact. The experiment itself was straightforward. For 2 months, he grew grasses from inside and outside the exclosure separately in pots of soil from either inside or outside. The results were dramatic.

"I get a little tingling down my spine when I think about it," he laughs. The grasses from both sides of the fence did much better— meaning a 34 percent increase in aboveground growth—when they were grown with the soil community from the grazed areas outside the fence.[28] "This is exciting because it suggests that somehow grazers are changing the composition of the microbial community, and this community is functionally different because it's facilitating plant growth."

But what was different about the community? Frank treated some of the pots with fungicide and the differences in plant growth disappeared, suggesting that the fungal community was responsible for the effect. Specifically, he believes the arbuscular mycorrhizal (AM) fungal community in the grazed areas may somehow provide more effective symbiotic partners for the grasses than the AM fungi on the roots of ungrazed grasses. Indeed, collaborator Catherine Gehring from Northern Arizona University extracted AM fungal spores from the soil and found that the abundance and diversity of spores were greater in the grazed soils and that the identity of the players in the spore community was also different. (Recall from the previous chapter that AM fungi are the most widespread of the mycorrhizal symbionts. They are much harder to study than ectomycorrhizae because they do not form mushrooms and their hyphae form branching structures called "arbuscules" within the root cells of the host plant rather than forming visible structures on the root surface.)

Frank believes that since grazing spurs root production and carbon dumping by Yellowstone grasses, this may also increase the "investment" the grasses make in the mycorrhizal partnership. And the AM fungi in turn may help the grasses make faster and better use of the nitrogen being liberated by the enhanced bacterial activity around their roots.[29] The story is still unfolding, however. An earlier study found that grazing actually disrupts the mycorrhizal partnership.[30]

Despite the decisive effect of the fungicide on his pot experiments, Frank is not ready to write off the potential for important changes in the bacterial community in grazed grasslands, too. Wiping out the fungi might have eliminated closely associated bacteria that also have a hand in boosting plant productivity.[31]

More than a year later, I learn that one of Frank's graduate students, Stacey Massulik, and Syracuse soil microbial ecologist Andria Costello are using genetic probes to determine the richness and composition of the bacterial community inside and outside a number of exclosures. They are confirming that the composition of the bacterial community is indeed different in grazed and ungrazed areas, although the diversity of bacterial species is similar. More than half the bacterial species are found in either the grazed community or the ungrazed community, but not in both. Another of Frank's students, Tanya Murray, is finding evidence that grazers also have strong effects on the composition of the mycorrhizal community, although the diversity of AM fungi appears unaffected.

This work comes at a time when a growing number of researchers around the world are investigating the ecological consequences of changes in the identity or biodiversity of the living communities underground. Fungal disease agents in the soil, for instance, can play a powerful role in shaping the aboveground plant community by weakening their hosts and permitting their replacement by or coexistence with other plants.[32] The identity and talents of mycorrhizal fungi in the soil community can also affect the diversity and productivity of the plant community, as well as the competition among plant species.[33] Many other soil denizens, from microbes to nematodes, may also influence the makeup and workings of the aboveground community, through either negative or positive feedbacks.[34]

Clearly, large grazing animals in Yellowstone, with the interaction and cooperation of grasses and soil creatures, influence energy (carbon) and nutrient flows in a way that has made these grasslands sustainable over time. But that doesn't mean these ecosystems or others can't be disrupted.

"You can degrade any system, drive it down, if grazing intensities are too great," Frank tells me. "But at the current rates of herbivory, that's not what we're seeing here. We're seeing stimulation."

I ask how managers could monitor or judge whether grazing had reached its limit.

"Reduced productivity would be a good indicator of overgrazing here, although measuring production in a grazed grassland is very time-consuming and expensive," he says. Certain types of human management practices are likely to decouple grazers from coevolved feedbacks with plants and soil. Herding and fencing of animals, for example, alter the natural migration patterns and the seasonal rhythm and timing of grazing on a range. Stocking cattle and sheep at high densities with the help of predator control, supplemental food and water, and veterinary care can also disrupt the natural feedbacks between grazers and the plant and soil communities. Ironically, without the natural intensity and timing of grazing to stimulate productivity, grazing-tolerant grasslands such as Yellowstone and the Serengeti may actually decline in carrying capacity and support fewer animals.[35]

The message from Yellowstone is not that grazing is "good" or "bad" for the range, but that we cannot understand how ecosystems will respond to grazing and other disturbances without considering what is happening below as well as above the ground. Plant-eating animals interact in complex ways with other forces, from climate and geology to plants and soil creatures, to shape ecological processes in grasslands, forests, and other systems. Exclosure experiments involving reindeer in northern Scandinavia, exotic deer in New Zealand forests, sheep and deer in the Scottish Highlands, moose in the boreal forests of Michigan, and many other habitats around the world show that grazers can have positive, negative, or even neutral effects on plant productivity and nutrient cycling.[36]

On the high arctic island of Spitsbergen, for instance, wild reindeer promote nutrient cycling and grass growth on the moss-dominated tundra largely via the dung they deposit. Richard Bardgett of Lancaster University in the United Kingdom and his colleagues added reindeer dung pats to some plots one summer and watched for three years before the impacts began to appear. By then, the patches where the dung had been added boasted a greater standing crop of grasses and enhanced microbial biomass in the soil. Also, the moss layer on these sites was significantly reduced, probably because of the enhanced activity of microbial decomposers.[37]

In the tallgrass prairies of the North American heartland grazing bison can spur positive feedbacks that increase plant species diversity, alter patterns of plant productivity, and speed up nitrogen cycling, apparently counterbalancing the effect of prairie fires and other stresses.[38] Yet there are many instances around the world where large grazing animals reduce the productivity and diversity of plant communities, including a number of arid grasslands where heavy grazing has led to the spread of shrubs and trees at the expense of the grasses.[39]

In forests, where trees live long and grow relatively slowly, nutrients also cycle more slowly than in grasslands, and grazers seldom seem to enhance aboveground productivity. Consider the moose that browse in the boreal forests on Isle Royale National Park in Michigan. Their preference for browsing on tender, palatable willows, aspen, and other young hardwoods retards the growth and abundance of these species and speeds the succession of spruce, balsam fir, and other conifers. The replacement of hardwoods by conifers reduces the amount of easily decomposed hardwood litter and leaves soil microbes to make their living by processing needles and detritus that are nitrogen-poor and laden with lignin, resin, and other recalcitrant compounds. John Pastor of the University of Minnesota and his colleagues found that, unlike in Yellowstone, soils in areas where moose browse are lower in microbial biomass and nitrogen, and this low fertility further suppresses the growth of the hardwoods that are the animals' favorite food plants. Only when fires or windstorms destroy large areas of spruce do the hardwoods get a chance to recolonize, speed up the nutrient cycle, and restart succession.[40]

There are many other instances where the selective appetites of animals alter the growth of their preferred food plants, and as a result, either accelerate or retard plant succession. Over time, of course, such shifts in the plant community alter the amount, timing, and nutritional quality of resources that succor the soil community, and this fosters changes belowground. Excessive grazing, such as insect outbreaks that severely defoliate trees, may also induce some plants to load their tissues with unpalatable phenolics or other noxious defensive compounds that discourage further attack. This is likely to make their litter more difficult to decompose, as well, and slow the nutrient cycle, although little research has focused on this.[41]

In the native rain forests of New Zealand, which hosted no grazing mammals until Europeans arrived in the 18th century, ecologist

David Wardle has been able to examine the impacts of introduced red deer and feral goats on the soil by making use of 30 fenced exclosures erected decades ago. Centuries of grazing by red deer, goats, and a suite of other exotic mammals from elk to wallabies have stripped the most nutritious and palatable plants from the forest understory in much of the country and promoted their replacement by unpalatable plants such as Blechnum ferns that grow up to 5 feet tall. Yet Wardle, who divides his time between Landcare Research in New Zealand and the Swedish University of Agricultural Sciences in Umea, has found no consistent impacts of this shift in plant resources on microbial communities and soil processes from one site to another. Strong positive effects are as common as strong negative effects, indicating that local conditions often influence the soil response to grazing. Densities of larger soil animals, however, from springtails and mites to spiders and beetles, have consistently declined in grazed areas, perhaps because of trampling.[42]

"The question for the long term is what happens to the next generation of trees?" Wardle told me. "New Zealand trees live a thousand years or more, so some of the trees alive now once had moas (giant flightless birds hunted to extinction a few centuries ago) browsing under them, then nothing for 500 years, then deer for 230 years, and feral goats for the last 100 years. When the current canopy dies in several hundred years, these soil changes caused by herbivores will probably affect tree succession."

Wardle and Richard Bardgett have examined similar idiosyncratic impacts of grazers around the world and come up with a prediction for when the effects on soil communities and ecological processes are likely to be positive or negative.[43] Negative effects of grazing on belowground processes are most common in infertile, low productivity systems with low herbivory rates such as boreal forests, the two predict.

On the other hand, positive stimulation by grazing is most likely to occur in highly fertile ecosystems with high grazing rates such as the Serengeti and Yellowstone, where more than half the greenery produced is consumed by grazers. When positive effects occur, they probably result from the kind of ramped up activity in the rhizosphere that we've just been talking about in Yellowstone. This activity is spurred when nutrients are returned to the soil in easily decomposed forms such as dung and urine, root exudates, or nitrogen-rich plant litter. This bounty stimulates microbial activity and reinforces soil fertility, and this feeds back to boost plant nitrogen uptake, photosynthesis, and productivity.

Globally, it turns out, sites of high soil fertility are also hotspots for herbivore diversity.[44] Thus, soil creatures that help green up the grasses of the earth's most productive rangelands are probably indispensable partners in helping to conserve the diversity of charismatic consumers such as elk, bison, zebras, wildebeest, and elephants. The belowground community is also a force we must consider as we humans attempt to sustain food production on ever-larger swaths of the global grazing lands.

IX

Restoring Power to the Soil

S oil remembers, and few memories persist more tenaciously than the changes wrought by agriculture. Especially intensive agriculture. Over the past century, farmers in the world's temperate climate regions have intensified production by sowing single crop species such as wheat or corn across vast acreages and adopting management practices that essentially decouple the plants from their dependence on the soil.[1] At the extreme, farmers in developed nations homogenize the soil by plowing, fumigate or sterilize it, protect and nourish the crop with pesticides and fertilizers, irrigate the crop, then harvest most of the grain and greenery and plow under the stubble, leaving the soil bare until the next season. The soil itself does little more than prop up the plants.

Such intensive cultivation spurs profound changes below-ground.[2] In natural systems, the rain of dead leaves, seeds, and other plant litter, as well as carbon-rich compounds exuded by roots, sustains a complex web of decomposers, grazers, and predators. When harvest and tillage (plowing) eliminate most plant debris from the surface, soil creatures such as beetles, millipedes, earthworms, and fungi that shred or process litter disappear.[3] Below the surface, bacteria quickly decompose the plowed-under stubble, exhaling the carbon as carbon dioxide and contributing little to the buildup of complex, long-lasting organic matter known as humus. Soil carbon stocks—the energy source that drives the soil food web—may decline by as much as 60 percent after natural ecosystems in temperate regions are converted to cropland.[4] What's more, wheat, corn, and other domesticated members of the grass family, unlike their wild cousins, have been bred to funnel most of their carbon into harvestable grains rather than to roots and root exudates that succor creatures in the neighborhood adjacent to the roots, known as the rhizosphere. Homeless, underfed, and out of work, an array of creatures from mycorrhizal fungi and fungal-feeding nematodes to pest-eating spiders drop out, leaving behind an impoverished, bacterial-dominated food web. Similar changes in the soil community, including a shift in power from fungi to bacteria, as well as a loss of plant diversity, take place when natural grasslands are fertilized to create "improved" pastures for livestock.[5]

Reversing these belowground changes is seldom as straightforward as ceasing to plow, plant, and fertilize. It can take decades or even centuries to recover the soil organic carbon stocks—a combination of plant litter and exudates, decay-resistant humus, and living organisms —depleted by cultivation. For many years after cultivation ends, the makeup of microbial communities in abandoned fields remains significantly different from that in never-plowed fields nearby.[6] Likewise, soil creatures from earthworms to fungal-feeding mites have proven slow to rebound even decades after offending land uses cease.[7] In turn, the continued absence of influential or "keystone" underground organisms such as mycorrhizal fungi, nematodes, and earthworms can leave the vital processes of decay and nutrient cycling profoundly altered.[8] And the lack of a complex web of consumers and predators reduces natural checks on disease and pest organisms in the soil.

The big question, of course, is "so what?" Do these changes in soil life and soil services really matter? If land remains in intensive agriculture, farmers can often maintain crop production levels for a time by plowing, fertilizing, and continuing to override soil services. This typically creates "leaky" systems that discharge excess nitrogen into streams, lakes, and coastal waters, however. In developed countries, where farmers are under increasing pressure to reduce water pollution, minimize pesticide use, and restore nature-friendly conditions on the farm, revitalizing the power and functioning of the soil community is essential. The same is true for subsistence farmers in developing regions who cannot afford to boost their harvests with chemicals in the first place.

The problem of soil degradation is vast. As noted in the opening chapter, more than 40 percent of the earth's plant-covered lands are considered degraded to some extent, including crop- and pasturelands damaged by farming and grazing practices that lead to soil erosion, nutrient depletion, salt buildup, or desertification.[9] The United Nations Food and Agriculture Organization (FAO) considers 70 percent of rangelands, 40 percent of rain-fed croplands, and 30 percent of irrigated croplands around the world degraded. In Africa, the result of declining soil fertility has been an 8 percent average loss in crop yields, with up to 50 percent losses in some regions.[10] Even in Europe, the FAO estimates that more than 540 million acres—an area four times the size of France—are moderately or severely degraded.[11] In the United States, improvements in tillage practices in recent decades have greatly reduced erosion rates; however, the Soil and Water Conservation Society considers that 30 percent of the nation's cropland (about 112 million acres) is subject to excessive erosion. Urbanization and salt buildup caused by irrigation also threaten significant areas of U.S. cropland.[12] Indeed, despite the spread of soil conservation efforts, two soil historians write, "in global terms the past 60 years have brought human-induced soil erosion and the destruction of soil ecosystems to unprecedented levels."[13]

The legacies of cultivation may also present a major obstacle to burgeoning efforts to restore species-rich natural ecosystems on abandoned crop- and pastureland. Many restoration efforts on fallow land across Europe, for example, have stalled or failed over the past decade, and soil ecologists suspect that the deeply embedded legacies of crop and pasture management are holding up recovery. In the tropics, meanwhile, a growing number of soil ecologists and agronomists

Intensive plowing, monoculture cropping, and chemical use break the natural bonds between plants and the soil community and spur profound changes belowground.

are hoping to harness the natural fertility-renewing services of underground communities to maintain production on cultivated lands, increase food security, and reduce pressures to clear more tropical forests for cropland.

Many of the most cherished features of the British countryside have fallen to the plow and fertilizer over the past half-century. Simon Mortimer, assistant director of the Centre for Agri-Environmental Research (CAER) at the University of Reading, points out a remnant of one of these features as we stand on the gently sloping lawn of a historic manor house at Bradenham village in the Chiltern Hills northwest of London. Mortimer directs my attention to a band of grassy fields running along the hill slopes across the valley, fields bisected here and there by hedgerows. At this distance, especially after the recent summer mowing, all of the fields look alike to me. Yet one of these fields hosts an increasingly rare habitat that hikers and birdwatchers eagerly seek out on weekends in the countryside: unfertilized, unplowed chalk grassland.

Chalk is a soft limestone, and the soils that develop on it are naturally alkaline and nutrient-poor, traditionally marginal for row crops. But infertile soils typically allow the coexistence of a richer diversity of plant life than fertile soils. Most natural ecosystems throughout the world have limited supplies of biologically available nitrogen, and many native plant species are adapted to function best in such low-nitrogen settings. When fertilizers or nitrogen pollutants are introduced to these ecosystems, the few plant species adapted to take advantage of high nitrogen levels grow tall and dense, outcompeting their neighbors and eventually reducing the diversity of plant species on a site.[14] The chalk grasslands, or chalk downs, Mortimer explains, are Britain's "tropical rain forest equivalent" because of the diversity of wildflowers, beetles, butterflies, and other nature they support. On pristine chalk downs, you may find 40 or 50 plant species per square yard, including a variety of orchids, rock rose, and wild thyme, and flitting among them, skylarks and stunning Adonis blue butterflies.[15]

The line of fields across the valley from Bradenham Manor ends abruptly at a dark green expanse of trees atop the rolling hills. These are beech "hangers"—overhanging beech woodlands—many planted in the 1870s when this area hosted a thriving furniture industry. Although I used the word "pristine" just above, both the highly prized beech woodlands and the remnant chalk downs are seminatural habitats.

"Three thousand years ago this whole valley would have been wooded with oak and ash," Mortimer says. "Most of southern England would have been wooded. There are very few circumstances in which we get natural grasslands because we don't have large native herbivores to prevent succession [the natural development and turnover of the plant community over time] to scrub and woodlands." The chalk uplands of southern England were cleared of trees in prehistory and have been maintained in a grassland state through the centuries by sheep grazing. Similarly, in the richer soils of the lowlands, the most valued landscapes are wildflower-rich Lammas Day meadows, named for the medieval schedule of communal grazing and hay cutting by which they were managed.[16]

Throughout Europe, centuries of traditional mixed farming and grazing have created similar versions of these relatively nature-rich grasslands. Yet over the past half-century, all have been vanishing as agriculture intensified.

"We now have only 5 percent of the flower-rich meadows and 20 percent of the chalk grasslands we had before World War II," Mortimer explains. The vast majority of hedgerows, ditches, ponds, and other rural features that impede the movement of large farm machinery have disappeared, too. The shortages Britain endured during the war spurred an all-out drive by the government to intensify farming and reduce dependence on imported food. In the 1970s, Britain joined with western Europe in adopting a system of production subsidies that prompted farmers to cultivate most of the remaining grasslands and meadows and quickly created massive food surpluses across the continent.

"With improved technology and machinery after World War II, and EU [European Union] price supports, farmers have been able to plow these chalk slopes and buy cheap fertilizer to grow cereal crops, mainly wheat and barley," Mortimer says. "So the farmed fields were pushed up right against the woodland edges." Farmers who specialized in row crops sold off their livestock, leaving steeper slopes ungrazed and open to invasion by rank (excessively vigorous in growth) grasses and scrub. Other farmers fertilized their downs and meadows and intensified livestock production.

These rapid and very visible changes in the rural landscape, along with parallel declines in farmland birds such as skylarks and gray partridges, soon sparked a public outcry. By the early 1980s, movements to "save our countryside" took hold. The British government began offering a variety of agri-environment schemes that pay farmers for reducing chemical use, maintaining hedgerows, sowing native plants along field margins, altering planting and harvesting patterns to allow skylarks to nest in the fields, or restoring grazing management of chalk grasslands. Other forces at work across Europe are now accelerating this shift away from intensification and toward more environmentally friendly forms of management that, for instance, reduce chemical inputs and maintain wildlife habitat on a substantial minority of farms. Among these forces are trade negotiations that have prompted the EU to begin backing away from guaranteed price supports for farmers that encourage overproduction. In addition, the adoption of the Convention on Biological Diversity (Biodiversity Treaty) in 1992 obligated member nations to take active measures to conserve native species and habitats. A pivotal event in farmland consciousness occurred in 2001 when an epidemic of foot-and-mouth disease broke out in Britain, leading to the slaughter of 4 million animals, the loss of export markets, and quarantines throughout the countryside. The quarantines were a particularly striking blow to rural businesses dependent on city dwellers fond of "rambles" in the countryside or family visits to "country life" parks where children can feed lambs and walk goats.

"The foot-and-mouth crisis brought recognition from policymakers that farmers don't just produce crops," Mortimer says. "They 'produce landscape' for cultural values, recreation, tourism, wildlife. Ten percent of farmland is now in some kind of agreement to manage in some way for the conservation of biodiversity, landscapes, or historic features."

The CAER research team and others are monitoring the results of many of these agreements to help policymakers target the subsidies more effectively. Many scientists and conservationists remain skeptical of agri-environment schemes, however. Even after two decades of subsidies for presumably nature-friendly management changes, for example, farmland birds in Britain continue to decline, and butterflies are disappearing even faster than birds or plants.[17] The EU has invested more than 24 billion euros (US$31 billion) since 1994 in a range of agri-environment schemes, yet some analyses of the results in the Netherlands and elsewhere in Europe have been highly critical.[18]

In addition to encouraging management changes on working farms, some governments now pay farmers to take land out of production and return it to nature. In the Netherlands, the government has emphasized buying up land from farmers. In Britain, landowners can receive payments for converting arable land back to grasslands. This conversion is now under way on a number of fields at the Bradenham estate, and that's why Mortimer has brought me out here this day.

Bradenham and the neighboring estate of West Wycombe belong to the National Trust, a charitable organization known more for its mission of preserving historic buildings, paintings, and furniture than its role as one of Britain's largest landowners. Mortimer shows me a small map of the Bradenham holdings. The few disconnected remnants of species-rich grassland on the slopes are outlined in green, improved (fertilized) grassland in dotted green, and cropped lands destined for restoration in red. Most of the species-rich grassland remnants are small, perhaps 5 acres each, and sited on the steeper slopes.

"The restoration sites were chosen where possible to link up with and extend the remnant grasslands," he explains, pointing out the scattering of red-outlined patches. "What the National Trust is trying to do now is keep the arable agricultural land in the valley bottom where the soils are deeper, but restore a band of chalk grassland on these steep slopes with shallow infertile soils between the beech woods and the arable land."

A great deal of money is being invested in such habitat restoration efforts, and many hopes and obligations are riding on the outcomes. Right now, however, ecologists cannot predict which efforts are likely to succeed. At most sites, fields are simply abandoned after the final harvest in hopes that valued native plants will recolonize. Letting nature take its course often fails, however. Even on sites where some grasses and flowering plants have returned naturally or been sown from seed mixes, recovery has seldom proceeded quickly. Researchers such as Valerie Brown, then director of CAER, estimate that with a hands-off approach, it could take a century or more for plant communities on former cropland to begin to resemble those on unplowed chalk downs.[19] With native plants and wildlife already in trouble, however, no one wants to wait a century. Brown, Mortimer, and colleagues across Europe believe that speeding up the recovery process will require paying more attention to life underground.

"Our studies have shown an incredible range in the value of what's been produced by 'arable reversion' schemes," Mortimer says. "From weed patches with thistles to sites quite close to the target. And we've been trying to work out why."

Mortimer, Brown, and research fellow Clare Lawson have examined 30 restoration sites in chalklands across southern England, half of them rated successes and half failures by government project officers. The effort has produced no checklist for success, but the soil and the neighborhood provide strong clues to a site's fate. Mortimer had shown me one of the successes, a "Rolls Royce site" at Aston Rowant National Nature Reserve along the edge of the chalk scarp west of Bradenham. Only 6 years after the final barley harvest, we were able to spot the small purple flowers of a Chiltern gentian, a rare chalk species that had established itself in the turf. The seeds must have come from the adjoining field, a remnant of unplowed grassland in the original reserve. "That's the key," Mortimer had explained. "The seeds are very close, and the soil is very shallow and infertile."

The key harbinger of restoration success, however, may be the abundance of specific types of creatures in the soil community. The team is hoping soil biology will eventually provide them with indicators that can be measured before land is entered into restoration programs. Already, for instance, they have found that the nematode community is significantly different between "good" and "bad" restoration sites: plant-parasitic nematodes dominate the failed sites while the soils of successful sites harbor a diverse array of fungal and bacterial grazers and omnivorous nematodes as well as plant-feeding species. And successful sites probably also have higher mycorrhizal biomass.

But are these differences a cause or a consequence of success or failure aboveground? Does the diminished soil life left by the plow stall development of a diverse natural plant community or vice versa? Clearly plants and soil creatures are mutually dependent in natural ecosystems, and the push-pull between the recovering communities above- and belowground must influence vegetation succession in these abandoned fields. Increasingly, soil ecologists across Europe believe

a lag in soil recovery is contributing to the failure of grassland restoration efforts. Brown and Mortimer at CAER have teamed up with Wim Van der Putten of the Netherlands Institute of Ecology and colleagues in Sweden, Switzerland, Spain, and the Czech Republic in a series of EU-funded projects to find ways to improve restoration strategies.[20]

"Something that needs more attention is the enormous time lag you get in recolonization by soil fauna if you leave it to natural recolonization," Val Brown had told me earlier. "Generally the soil fauna and certainly the microbes have much reduced mobility, and that's one reason they take longer to build up a population. So I think we need active intervention belowground as well as aboveground. You may get a quick fix on your wildflowers and your butterflies, but if the engine room of the soil isn't fixed, eventually your restoration will just collapse."

Very likely, a rich supporting cast of fungi and nematodes, mites and earthworms is needed to sustain gentians and orchids, skylarks and Adonis blues. Van der Putten agrees that this underground cast of characters may need a kick-start. Enhancing restoration is the practical goal, but part of the appeal of this program to researchers is the opportunity to probe the fundamental ecological processes of succession and community development, which in the long run could help to improve restoration strategies.

"For me, I want to see what drives succession and whether you can skip stages," Van der Putten tells me later. "It looks like a system has to go through every stage, but is this really necessary? Is there a critical factor that doesn't allow a system to jump from A to C without going through B? Maybe you can go directly to stage C by sowing the plant community, or maybe you also need to change something in the soil so the soil community is going from stage A to C, too. If the plant community is developing faster than the soil community, then after a while the belowground may be pulling back, preventing any further changes. We want to come up with some way to give it a tactical knock and bring the soil community along."

But what kind of "tactical knock" can accelerate the development of complex soil communities? Will increasing the diversity of plant species on a site do it? Or identifying and sowing specific native plants that exert a strong influence belowground, say by boosting mycorrhizal fungi? Or using only seeds of local origin? Or directly inoculating a site with soils and soil life transplanted from target natural areas? Or—because a richer array of native plant species can usually coexist on less fertile soils—reducing soil fertility by scraping off the topsoil, mowing and removing the plant litter, or instead, spreading sugar or sawdust to fuel a microbial population boom that ties up excess nutrients?

All of these ideas are being tested at project sites throughout the United Kingdom and elsewhere in Europe, including in a field that lies just north of the manor house at Bradenham. There, tall, dense vegetation in a checkerboard of 30 × 30 foot patches makes the study plots easily distinguishable amid the mowed pathways. The growing season is nearly over, and many of the tall brown stalks are topped with dried seed heads. The plots are strung out along the top of the slope below a fence that follows the woodland edge, and a half dozen students are scattered among them collecting vegetation samples and soil cores.

The oldest plots here and at the other sites in continental Europe were first set up in 1996. Seeds of either a high- or low-diversity mix of native grassland species were sown into the stubble of the final grain crop on some plots, and others were left unsown to colonize naturally. A remnant chalk grassland lies 200 yards away from where we stand, providing a potential seed source. The experiments were designed to address a number of hot topics in ecology that are relevant to restoration. One question is whether sowing a higher diversity of plant species helps suppress the seeds of arable weeds such as ragwort and thistles that often invade abandoned fields and retard succession to more diverse plant communities.

The teams found that higher diversity plant mixes did indeed suppress weeds and other natural colonizers more effectively than the average low-diversity mix, but the identity of the species mattered more than the number. If a low-diversity mix included one or two highly productive plants—for instance, fescue in the Netherlands—

that suppressed weedy invaders just as effectively as the richer mix.[21] More to the point here, the researchers also examined whether manipulating plant diversity can alter the development of the soil community. Their hypothesis was that a greater diversity of plants would result in greater biomass production, and the abundance of greenery, roots, and litter would stimulate a boom among decomposer microbes and soil animals. Contrary to expectations, however, a richer array of plants did not yield more green matter in these studies. Plant diversity also didn't seem to have any consistent effects on soil microbes, nematodes, mites, springtails, and earthworms, at least over the first 3 years.[22] Though the soil communities didn't seem initially to respond to what was happening in the plant community, the researchers expect that to change over time.

"I think for soil microbes, dispersal limitation probably isn't a problem and that it's just a question of conditions becoming more appropriate through time," Mortimer says. In other words, the decomposer bacteria and fungi are always in the fields at some level, or they disperse fairly easily through natural means such as blowing dust, and their populations will boom when conditions become favorable. "But I think for some of the soil macrofauna such as the earthworms it may be a different story. The diversity here may take a long time to resemble that of the established chalk grassland." The same may be true for mycorrhizal fungi.

To try to cut that lag time, the teams have been experimenting with direct efforts to recolonize the soil. In between the older plots, Mortimer shows me a series of "stepping-stone" plots—6 × 6 foot squares into which loose soil or 1-foot cubes of turf from the remnant grassland nearby have been added.

I ask how much soil they added, imagining trucks laden with new topsoil.

"It wasn't civil engineering, only a wheelbarrow job," he laughs.

Some of the new plots have also received a scattering of seed-filled hay harvested from the same grassland, and the transplanted soil itself carries a native seed bank. The soil additions have greatly accelerated plant community development on the stepping-stone plots,

Mortimer says. Plantwise, they now resemble 15- to 30-year-old chalk grasslands rather than 5-year-old grasslands, and rare chalk species such as white-flowered eyebright have spread from these plots to colonize the rest of the field. But it remains to be seen whether the soil transplants are speeding up the dispersal of mycorrhizal fungi, nematodes, and other organisms from later successional soil communities.

Again, succession refers to the progressive replacement of one kind of plant and animal community with another. It is a directional process, beginning when vegetation on a site is damaged or destroyed by fire, flood, plowing, or other disturbance. Early successional plant species quickly colonize, altering the terms of life for animals above- and belowground that rely on plants for food or eat the animals that do. These early successional plant and animal communities then alter the environment for plants and creatures that arrive next, and so on, driving the turnover of mid- and later successional communities until disturbance restarts the process. Van der Putten and his team in the Netherlands are finding, in fact, that individual plant species can have a profound influence in shaping the soil communities around their roots, and these soil communities, in turn, can help drive turnover and succession in the plant community.

Only days after visiting Bradenham with Mortimer, I find myself in another small grassy field on which similar hopes are riding. This field lies on the grounds of an agricultural institute near the town of Wageningen, about 10 miles west of Arnhem in eastern Netherlands. I'm here with Wim Van der Putten, who heads a department called "Multitrophic Interactions" at the Netherlands Institute of Ecology nearby. "Multitrophic" means multiple levels of the food web above- and belowground, such as plants, the pathogens or herbivores that attack or consume their leaves or roots, the predators that eat these consumers, and so on. Like Mortimer and Brown, Van der Putten believes that restoring complex multitrophic interactions in the soil is critical to restoration of lands impoverished by agriculture. The grassland site we are visiting, however, has apparently stalled midway to that goal.

This fenced field entered restoration 20 years ago after a history as improved grassland—a more common restoration starting point in the Netherlands than abandoned cropland. Restoration efforts have focused on "cultivation management," which means annual mowing and hay removal to try to reduce the soil's fertility.

"This is a basic goal in grassland restoration, bringing down fertility," Van der Putten reminds me. "The productivity of improved grasslands will be 10 tons of dry matter per hectare [4 tons of hay per acre] per year. At 6 tons per hectare [2.4 tons per acre], it's supposed to become more species rich, according to diversity-production curves. But it hasn't reached that. This plot is down to about 7 tons [2.8 tons per acre] per year."

It may never reach target, however, because airborne nitrogen pollution dumps the equivalent of 36 pounds per acre of fertilizer on this field each year. "At a certain point, haying just keeps soil fertility from increasing, and the decline in fertility has plateaued," he says.

Nitrogen pollution from farms, factories, and automobiles is severe across northwestern Europe. As we saw earlier in this book, the rain of nitrogen has already caused declines in mycorrhizal fungi and other changes in the soil community in many parts of Europe.[23] Also, nitrogen-loving plant species have increased over the past half-century while nitrogen-intolerant species have decreased in grasslands, heathlands, peatlands, and forests. A recent survey of economically important grasslands in Britain—specifically, bent grass and fescue pastures that are also common in other parts of Europe as well as North America and Australia—confirmed that species richness drops as nitrogen deposition rises. Most vulnerable are plants such as eyebright, plantain, harebell, heather, and moor grass that are native to naturally infertile landscapes. Current average nitrogen deposition rates in Britain and central Europe are sufficient to cause a 23 percent loss of grassland plant diversity, the study indicates. In the eastern United States, it's enough to knock species diversity down 5.2 percent.[24]

The Netherlands is the most densely populated country in Europe and the third largest agricultural exporter in the world thanks to

intensive livestock operations and greenhouses and fields full of flowers and vegetables. Huge volumes of ammonia wafting into the air
from manure and fertilizers add to nitrogen emissions from cars and
industries.[25]

"Maybe it's just crazy that we've tried to give some areas back
to nature," Van der Putten says with little hint of discouragement.
"Maybe the atmospheric deposition here is not very suitable for getting these low productivity areas reestablished. We will always have
to mow intensively and graze them." In restoration efforts that begin
with the bare soil of abandoned croplands rather than improved pasture, the Dutch increasingly bulldoze off the top foot of nitrogen-rich
topsoil and leave nature to take its course. Too often, however, in all
of these restoration approaches, what nature comes up with first is an
unsightly field of thistles that angers farmers and the public.

"The initial stages of succession don't look nice to people because
you get these arable weeds," he explains. "And the farmers say, 'we
and our grandparents have been taking care of the land for years, and
now the nature conservation people get it and look what kind of a
mess they make out of it.'" It's a sentiment he understands: He and his
family live in the reed-thatched farmhouse his great-great-grandfather
bought in 1853, and he tends a small field of wheat along with an or-
chard and vegetable garden.

Van der Putten and his colleagues, of course, hope to find ways
to skip these early stages in succession or else push a faster turnover
toward the species-rich targets, and they are looking to soil management for ways to accomplish this. "Right now, restoration is pro-
ceeding just by vegetation management," he says. "But by not
managing the soil biota, people might have been overlooking something which is crucial for ecosystem development."

One attempt to manage soil biota is under way on the field we
are visiting today, which hosts "stepping-stone" plots just like those
at Bradenham. However, it turns out to be much more difficult to
transplant soil or patches of turf and seeds into the dense growth of
this still-fertile grassland than into the bare dirt of a fallow field. Van

der Putten points out the bamboo stakes that mark plot boundaries, which are otherwise hard to distinguish in the thick, low-mown sward (grass cover). Apparently the transplanted species also had difficulty finding an opening here.

"The main conclusion of these experiments is that most of these treatments failed," he says. "They couldn't enhance succession. So it seems that the closed sward is very important in slowing down the process of succession, even when you bring in later successional plants and soil organisms. It could very well be that you have to open up the vegetation [by tilling or removing patches of turf] in order to give it a push toward the new stage and give more opportunities for these later successional plant species to germinate."

If such plants do get a toehold, however, the soil community on this site is capable of favoring them and pushing succession toward a species-rich system. Van der Putten knows that because of the dramatic result that graduate student Gerlinde de Deyn got when she put plants and soil animals from this field—considered to be at the midpoint of succession after 20 years of restoration—and two other sites of different ages to the test in greenhouse microcosms.[26] The other sites included a recently abandoned pasture that represents early succession and a species-rich natural grassland that represents the restoration target. De Deyn took soil from this field, sterilized it to eliminate the soil community, filled 32 pots, and planted each with a mix of plant species from all three successional stages. After giving the plants 6 weeks to establish in the sterile soil, she inoculated each with a natural assemblage of soil animals—nematodes, mites, springtails, and wireworms (click beetle larvae)—from one of the three stages and let the plants and soil animals interact for a year in a greenhouse.

In all cases, the soil animals suppressed the dominant plant species from the early successional community and allowed the later successional plants to grow more abundant. The effect was strongest with soil animals taken from this mid-successional field. In containers where no soil animals were added, the mid-successional plants eventually dominated, blocking turnover and development toward the target late-stage plant community. De Deyn and her colleagues concluded that soil animals from all stages "profoundly enhance vegetation succession and the homogeneity of the plant community by reducing the biomass of the dominant plant species."

Just how did soil animals exert their influence? Apparently by nibbling selectively on the roots of the dominant plants. In the early successional soils, the root-feeders were mostly nematodes; in the mid- and later stage soils, wireworms joined the nematodes. The animals affected plant turnover indirectly by weakening the dominant plants and allowing subordinate plants to expand. That's much the same effect that aboveground grazers, from bison and moose to sheep and cattle, can exert. As we saw in the last chapter, grazers can strongly influence the composition and successional state of plant communities either by boosting or suppressing their preferred food plants and accelerating or retarding turnover of species. De Deyn's work provides some of the first clear evidence that tiny invertebrate animals belowground can do the same.

Of course, other factors such as nitrogen levels also play a role in driving succession, and de Deyn wondered how these factors might interact with the influence of soil animals. When she took pots of sterilized soil without animals and added various levels of fertilizer, a few grasses dominated and grew more prolific. In contrast, in pots with an intact soil community, the plant assemblage remained more diverse and grass dominance decreased with time even when low levels of fertilizer were added. Only with high fertilization did diversity decline.[27]

We've been talking about soil animals, but the identity of microbes in the soil, ranging from pathogens to mycorrhizal fungi, also profoundly influences which plant species can coexist in a community and how succession proceeds.[28] Microbial influence can be quite variable, however. Mycorrhizae, for instance, can enhance plant diversity directly by helping subordinate plants forage for nutrients and survive. On the other hand, mycorrhizae may promote a dominant species at the expense of diversity and succession.[29] Root pathogens, likewise, can suppress dominant plants or rare ones. In the early 1990s, Van der Putten unraveled a now-classic example of the power of belowground pathogens in the coastal dunes that harbor the Netherlands' richest plant life.

Marram grass is the main sand-stabilizing species in Dutch coastal dunes and the first plant to colonize new dunes. During the growing season, mycorrhizal fungi as well as plant-feeding nematodes and fungal pathogens all colonize the grass roots. In winter, while marram grass shoots are dormant, wind-driven sand accumulates and the dunes grow taller. In early spring, the grass sends up new roots into the fresh sand layer, escaping its pathogen complex for a time until these creatures, too, disperse upward and proliferate. The grass keeps growing like this, one step ahead of its pests, until the dunes get too tall or sand deposition declines. At that point, the grass falters and later successional plants such as sea buckthorn or elms take over.[30]

The marram grass story illustrates the fact that plants and their root symbionts face not one but an array of belowground pathogens and root-feeders at any given time. Interactions between plants, mycorrhizae, and multiple attackers have received little research attention, however. And only recently have researchers begun to probe what happens to plants when they are beset by insects and other herbivores aboveground at the same time their roots are under siege.[31] Effective management of succession, and thus restoration, depends on understanding such complex biological interactions, and that requires increasing collaboration and interaction at the human level. Adding mycorrhizal fungi to a study of soil insects, for instance, requires that an entomologist team up with a mycologist, since few traditionally trained scientists cross these two disciplines. And factoring in the impacts of earthworms, nematodes, springtails, or other soil creatures requires an even larger interdisciplinary team.[32]

What's more, the responses of plants themselves to assaults from above and below need to be factored in. Plants may mobilize defensive chemicals in their roots just as they do in their greenery to stave off attack; signal natural enemies to prey on their attackers; develop tolerance or resistance to some herbivores and pathogens; or devise strategies to "outrun" them, as marram grass does in the dunes.[33]

Although researchers are just beginning to unravel the complexities of soil-plant interactions, Van der Putten believes one message stands out: The nature of the soil community you start with largely determines whether the soil will boost or hinder the process.

"What Val [Brown] has found in chalk grasslands and I have found in dunes is that during succession, the soil community can speed up the replacement of plant species in the vegetation community," he says as we are leaving the grassland site. "But on arable lands, we think that to some extent, the soil community may just impair the development of the plant community." That's because former cropland enters restoration with a greatly diminished soil community, and the simplified structure of the soil food web apparently makes it difficult for soil organisms from more complex systems to colonize and make a living. There's a chicken-and-egg problem here: These creatures require a food web based on litter, root exudates, and other plant offerings, which may not be in place on fallow land. Yet the appropriate plants may not be able to establish until certain elements from a more complex soil community arrive.

And this brings us back to a point I mentioned earlier: Even if the number of plant species that occupy a site has no consistent impact on soil life, individual plant species can strongly influence the soil community around their roots. And these localized, species-specific influences eventually combine to shape the larger community both above and below the ground. Van der Putten and his colleagues have found these effects by sampling the soil beneath individual plants at a restoration site in the dry, sandy heath soils of the Veluwe region, where the Dutch government has been buying up farmland around Hoge Veluwe and Veluwezoom national parks. The site, in a former cornfield a 40-minute drive north of the grassland project, contains a series of experimental plots much like those at Bradenham, with plants growing in high- and low-diversity mixes and also in individual monocultures.

Ragwort was one of the first and most aggressive colonists in the plots, Van der Putten points out. Yet where ragwort invaded most densely, it also grew stunted and eventually declined, causing the researchers to suspect a buildup of soil pathogens under the plants. Indeed, when other researchers used molecular techniques to "fingerprint" the microbial community, they found a different fungal pattern under the stunted plants than under taller ragwort plants, suggesting the presence of fungal pathogens.[34]

Coring under another species, plantain, revealed that it appeared to be suppressing nematodes. Gerlinde de Deyn followed this up in an experimental field back at the institute, growing plantain and other species for 3 years in monocultures and in mixtures of up to 16 species. The plantain in her plots, too, harbored low nematode numbers, perhaps because plantain exudes defensive compounds that kill or repel them. In contrast, oxeye daisy hosted high numbers of nematodes. Nematode diversity varied more from one plant species to another than between different levels of plant diversity. Not surprisingly, plant-feeding nematodes and others that interact intimately with plants were most affected by the identity of the plants above them.[35]

"I wonder, what does it mean for the surrounding plants if a certain plant is suppressing or building up pathogens or even beneficial organisms," Van der Putten says. "We are interested in how vegetation patterns are driven and how biodiversity is maintained by local effects of the soil community. How succession is developing in relation to the presence or absence of certain soil organisms. And from our point of view, these small-scale effects are the first step toward the large-scale development of the plant community."

Van der Putten is entranced by the idea that an understanding of succession might allow us greater leeway in using the land again if necessary, but in a sustainable way. He compares this idea to the tradition of shifting cultivation in the tropics. "If you need to, you should be able to use the grassland for agricultural production again if there's trouble in the future, war or something," he muses. "The question is just, can we take a natural system and turn it to an arable system and then bring it back again sooner than nature would do it? That's the question that drives our work."

Shifting cultivation, swidden farming, or slash-and-burn agriculture has been practiced for many centuries in the tropical forests of Asia,

Africa, and South America. Traditionally, this type of farming involves clearing trees or scrub, cropping the land for a few seasons, then fallowing it and allowing elements of the forest to regenerate for a decade or two before the land is cleared and cropped again.[36] Escalating population growth and poverty, however, have sent hundreds of millions of desperate farmers to carve away at shrinking fragments of forests that once sustained only tens of millions of swidden farmers. More land is being cleared, worked longer, and then abandoned when the forest-derived soils—typically highly acidic and low in organic matter and nutrients—have been severely degraded. Although tropical rain forests are under siege by many forces, including commercial logging, cattle ranching, and gold mining, the greatest cause of deforestation throughout the tropics remains unsustainable slash-and-burn farming.[37]

Chemical fertilizers could extend crop production on exhausted tropical soils, of course, but few subsistence farmers can afford them. Instead, international organizations such as the Convention on Biological Diversity and the United Nations FAO are increasingly promoting solutions such as "soil biological management" that are designed to take advantage of the natural soil-renewing services of belowground biodiversity.[38] The goals of these efforts are to prevent the most debilitating effects of cultivation on tropical soils and their living communities, and as a result, increase food security and minimize the pressure to clear more forest.

But how much soil biodiversity, and what types of creatures, need to be maintained in a crop field, pasture, plantation, or other agricultural setting to sustain soil structure and fertility and prevent erosion? And what can a subsistence farmer do to maintain this biological richness without compromising his family's food supply?

Research relating soil life and its activities to the maintenance of soil fertility is sparse and has seldom been translated into practice on the farm. Agricultural research and development programs, as well as extension and education efforts, have long neglected soil biology, the FAO notes, but recently there has been a "move away from the conventional focus on overcoming soil chemical and physical constraints . . . to a focus on soil health through an approach centered on soil biological management. . . ."[39]

To help advance this approach, the United Nations Environment Programme in 2002 announced a 5-year, $26 million project called Conservation and Sustainable Management of Below-Ground Bio-diversity (BGBD).[40] Initially the project is targeting soil life in seven tropical countries: Brazil, Mexico, Ivory Coast, Uganda, Kenya, India, and Indonesia. Project teams in each country are surveying and in-ventorying soil constituents, identifying indicators that will help in monitoring their status, and testing alternative land management practices that will enhance conservation of belowground biodiver-sity.[41] BGBD is building on the work of another global partnership known as Alternatives to Slash-and-Burn (ASB), which was supported by a consortium of international development and agricultural re-search agencies.[42]

During the 1990s, ASB-sponsored research teams surveyed key groups of soil creatures along gradients of land use from intact forests to forest-derived fallows, fields, and pastures in Cameroon, Brazil, Peru, and Indonesia. Because most of the soil life biologists find in the tropics remains uncataloged and unnamed, the ASB teams targeted some of the better known groups, including earthworms, termites, ants, woodlice, millipedes, nematodes, arbuscular mycorrhizal fungi, and nitrogen-fixing bacteria. The researchers found that the identities and numbers of some key functional groups of soil animals—especially ecosystem engineers such as soil-feeding termites, earthworms, and ants—shift dramatically as land use intensifies, although not all groups respond the same way to land use change.[43]

David Bignell of Queen Mary, University of London, who led the ASB termite surveys, pointed out that the next step under BGBD is to link certain levels of termite loss or other shifts in soil animal popu-lations to specific management practices, and in turn, to detectable changes in soil qualities or crop yields. "The sampling transects are there now, and we can look at things like crop yields in relation to what's in the ground. The idea is to look at the crop production in the same place that you look at the soil fauna and then work out what the relationship is. And at that point, we should be able to say where the loss of soil biota critically impacts production."

The step beyond that is to find ways to prevent that loss, Bignell continued: "The aim of our project is to find management systems for subsistence farmers that reduce the rate at which they need to clear new land. And also make the existing land that they've got last longer, but without fertilizer. Fertilizer is the easy answer, but you can only turn to that if you're cash rich."

What might biodiversity-friendly management approaches include? I wondered. The presumption of BGBD is that agricultural *diversification* can counter or reverse the ill effects caused by agricultural *intensification*.[44] That means creating a mosaic of different land uses at different levels of intensification.

Bignell translated that into British landscape terms for me: "It looks like small is beautiful. If you're a farmer but your land mosaic includes refugia [habitat suitable for native plants and animals]—say hedgerows, little woodlands, a few ponds with a bit of vegetation around them—then you can sustain biodiversity even in the bits you farm. But if you have large fields of many hectares and the hedgerows are all taken away, then the biodiversity just goes down to nothing. It's the same thing in the tropics. If you just clear a little forest infrequently, then the biota will reestablish itself quickly and the land can be restored. There is a critical point somewhere, however, where the amount of land you clear is large—perhaps 5 or 6 hectares—where the biota can't really reestablish itself."

A landscape mosaic designed to sustain tropical soil biodiversity might include areas of mixed or multispecies cropping, crop rotations, buffer strips of vegetation at field edges to slow erosion, small fields, no-till approaches where seeds are planted amid the stubble of the previous crop, or an array of organic soil amendments. The idea is to manipulate soil life indirectly by providing more diverse habitats and resources to nurture soil biodiversity, improve nutrient cycling, and enhance natural pest and disease control.

This story is still unfolding as researchers throughout the tropics attempt to translate their ideas into practical management schemes that subsistence farmers can use. Perhaps the lessons from the tropics, as well as the search for effective agri-environment and restoration approaches in the developed world, will help us learn to restore power to the soil throughout the landscapes we rely on for food, fuel, and other essentials. As we've seen throughout this book, we ignore or degrade life underground at our own peril. Complex soil communities not only help to renew soil fertility but also dampen pest and disease outbreaks, nurture the growth of trees and other plants, improve water quality, regulate greenhouse gas emissions, and perform myriad other services vital to our well-being. We must begin to work with the soil, not against it, and learn how to let life underground work for us.

Epilogue

Our senses, wondrous as they are, limit our perceptions of life on earth. We and the whales, the frogs and the owls, the firs and the orchids are grand anomalies. Most life is microscopic, or at least inconspicuous, and it lives and works in darkness within the mud and dirt—the earth—of the Earth. Among these multitudes are the very microbes that terraformed a hostile young Earth (made the planet habitable) billions of years ago, and billions hence will be the last to go, long after the dying sun has seared away all surface life.[1] Until then, we rely on them and the rest of the teeming ranks of microbes and animals underground to maintain the earth in a life-friendly state.

The top few inches of the earth's crust, whether on land or under lakes, rivers, and seas, remain a largely unexplored wilderness, however. Only a few percent of organisms underground have been identified, and soil and sediment life has been virtually ignored in conservation research and policy.[2] Until recently, the same was true in ecology and even agronomy and soil science, and for that reason we still know comparatively little about how even the named creatures belowground make their living.

Over the past decade, however, research on soil biodiversity and soil ecology has exploded. Aided by advances in molecular genetics and other new techniques, growing numbers of scientists from a wide range of disciplines are pushing into what actually may be, at least on this planet, "the final frontier."[3] Researchers are probing who is down there, how they interact with one another and with life above, whether food webs and communities belowground follow the same rules as those above, and how the activities of soil creatures influence the ecological processes that help keep the water pure, the air breathable, the climate benign, and the surface of the earth green.

The attention that soil and the life within are finally beginning to receive is coming none too soon. The United Nations Environment Programme has declared the soil "the largest source of untapped life left on Earth."[4] Yet that life, just like numerous species aboveground, is increasingly threatened by our activities. Unique and wondrous life forms may be vanishing before we even learn of their existence.

Throughout this book, I've talked about legacies, both directly and indirectly. Soils and sediments themselves are a legacy of geology, climate, and life. Every plant, animal, and microbe adds its tiny mark to the soil, in life and in death. We humans, too, leave marks in the soil, often not very subtle ones. Instead, our activities too often leave an enduring legacy of degradation and diminished life underground. Some of those destructive activities have been featured in this book. But I have also tried to show how ecologists and land managers are using our burgeoning knowledge of soil life to try to change damaging practices and harness the power of life underground to restore and maintain the earth's lands and waters:

- Clearcutting can destroy much of the life of a forest, from bears and owls to the vast underground web of mycorrhizal fungi that serves as an indispensable lifeline between forests past and future. Soil ecologists are now teaming up with foresters to find new approaches to timber harvesting that will ensure both the preservation of forest biodiversity and the success of forest recovery.

- Intensive agricultural practices too often degrade soil life and decouple the dependence between plants and the soil, replacing natural pest control and nutrient cycling services with chemicals. Soil ecologists and agronomists are learning to rejuvenate and maintain complex underground communities, both to sustain the fertility and productivity of cultivated lands and to restore natural plant and animal life on abandoned fields.
- Much of the earth's grassland has been converted to pastures, and in many parts of the world, livestock has compacted the soil, stripped away plant cover, and left the land open to wind and water erosion. Ecologists are investigating how large herds of native grazing animals have coexisted with grasslands for millions of years without degrading them—a phenomenon that involves feedbacks between grazers, plants, and life underground —in hopes of learning to better manage our pastures and rangelands.
- Humans are increasingly altering native plant, animal, and soil communities, and often ecological processes, by introducing nonnative species into new habitats. By studying the impact of exotic earthworms on the flowers, trees, and animal life of invaded forests, ecologists are coming to understand the power of these tiny ecosystem engineers—a power that in other parts of the world is being harnessed to restore fertility and enhance crop production on degraded lands.
- Activities from fertilizer production to fossil fuel burning help feed and power the world's burgeoning human population, yet they also generate excess nitrogen that exacerbates acid rain, global warming, water pollution, and coastal dead zones. At the same time, we continue to destroy wetlands that harbor vast concentrations of microbes that can deactivate that excess nitrogen. Ecologists are working to clean up polluted rivers and reduce dead zones in coastal waters by restoring both wetlands and the denitrifying services that wetland microbes provide.

- Because oceans cover more than 70 percent of the globe, submerged sediments make up the most extensive ecosystem on earth and harbor one of its richest animal communities. As with soil creatures on land, life in undersea sediments is increasingly at risk from a variety of human activities, from nitrogen pollution to fishing practices such as bottom trawling and dredging. Creatures that burrow in sediment are vital to nutrient cycling in the oceans just as soil bioturbators are on land, and marine scientists are hoping to win protections for these bottom-dwellers before our assaults on them affect the entire ocean food web, from plankton to fish and whales.

All of these human actions and many others I have not explored—including paving roads and building sprawling expanses of shops and homes atop the soil—create impacts far more visible to us than are their effects on the soil and its creatures. Indeed, we usually take notice only when changes belowground set in motion a cascade of unwanted consequences aboveground, such as faltering crop yields, disappearance of beloved wildflowers, or failures of cutover forests to regrow.

My intent, however, is not simply to criticize our treatment of the earth, but rather to illustrate how profoundly the creatures of mud and dirt shape the world we experience. The growing interest in life underground comes not just from ecologists and taxonomists but also from foresters, agricultural researchers, marine biologists, range scientists, and others who recognize that understanding the influence of soil and sediment creatures—and in turn, our impact on them—is vital to the future productivity and health of our lands and waters.

There are many more tales that could be told about the involvement of soil and sediment biodiversity in issues of global concern. For example, the buildup of carbon dioxide and other greenhouse gases in the atmosphere is helping to drive changes in global climate. These global changes are also affecting plant growth and, indirectly, soil life. Soils store massive amounts of carbon, and the activities of belowground creatures determine whether carbon is stored long term or recycled through decomposition to be taken up by plants and animals and respired to the atmosphere. A critical question is whether indirect, climate-driven changes in belowground life and activities will speed or slow the release of soil carbon and thus, feed back to accelerate or slow further warming. Warming in northern latitudes, for example, may speed microbial decomposition and release of carbon long stored in peat and frozen soils. Yet the faster turnover of nutrients may also spur increased plant growth that draws in more carbon.[5]

Another example of the involvement of soil life in issues of global concern is the potential environmental impacts of planting genetically modified (transgenic) crops and trees or releasing genetically engineered microbes. The most widely planted transgenic crops carry herbicide-tolerance genes that allow farmers to spray more weed-killers on their fields. Other plants carry genes from microbes that enable them to produce toxins (Bt) intended to kill crop pests. These toxins may show up in the soil food chain through root exudates and plant litter, with unknown effects on the soil community and, in turn, unpredictable feedbacks aboveground.[6]

The intersection of soil and sediment life with high-profile issues such as these has been a Cinderella tale, attracting for so long meager funding and scientific interest. I've been pleased to learn in my travels and research how quickly this is changing. Rather than adding to our burden of environmental concerns, growing understanding of the underground world can help us to redefine and clarify our stewardship. Most of the threats to soil and sediment life come from processes and activities that we already recognize as destructive or unsustainable. Understanding life underground can help us to devise more effective responses to global concerns ranging from eutrophication and land degradation to climate change.

The first time I met Diana Wall, I asked her what message she wanted to send to people about life in the soil. "I want them to think whenever they walk," she replied. "There's a whole world under there. I want them to step lightly."

Stepping lightly could mean creating refugia or reserves, just as the Antarctic Treaty protects the valley that Wall thinks of, only half in jest, as "Nematode National Park." Recently, researchers at the University of California, Berkeley, called for protections for more than 500 rare and endangered soil "series"—each equivalent to a plant or animal species—among the 13,000-plus soil series in the United States.[7] Protecting rare soil types from pavement and plow would also preserve the unique communities that rely on them, both aboveground and below.

Edward O. Wilson endorses the creation of "microreserves" to protect "microwildernesses" and the tiny life within, although he does not view them as substitutes for full-scale biodiversity reserves: "People can acquire an appreciation for savage carnivorous nematodes and shape-shifting rotifers in a drop of pond water, but they need life on the larger scale to which the human intellect and emotion most naturally respond," he writes.[8]

Conserving belowground biodiversity for its intrinsic value alone is a worthwhile goal. Beyond that, soil creatures also represent a treasure trove of millions of unexplored genomes that may yield new antibiotics and pharmaceuticals, novel genes and enzymes, and microbes with talents that could be put to work attacking pests and pathogens, or cleaning up pollutants.

Such values could be protected in a series of small reserves off limits to human activities, but soils and sediments supply too many essential services for the life within to be reduced to the status of zoo or boutique specimens. We require robust, abundant, free-living populations of creatures working underground everywhere, from swamps and deserts to forests, grassy plains, and abyssal mud. Across most of the earth's surface, stepping lightly must mean developing sustainable ways to manage and continue to enjoy the ecological life support services that belowground communities provide. Most of us will never respond to microbes or nematodes with the emotional connection we muster for elephants and eagles, but we should at least acknowledge that we and the charismatic surface creatures we value would be doomed without them. My hope is that we can learn to step, not only lightly, but also with wonder and awareness of the world underground.

Notes

I. INTRODUCTION: Opening the Black Box

1. Kerr, R. A. 2004. Opportunity tells a salty tale. *Science* 303: 1957.

2. *Merriam-Webster's Collegiate Dictionary, Eleventh Edition.* 2003. Springfield, MA: Merriam-Webster, Inc.

3. Richter, D. D. and D. Markewitz. 1995. How deep is soil? *Bio-Science* 45: 600–609. Quote p. 600.

4. Usher, M. B. 1985. Population and community dynamics in the soil ecosystem. In *Ecological Interactions on Soil: Plants, Microbes and Animals*, ed. A. H. Fitter, 243–265. Oxford: Blackwell Scientific.

5. Brussaard, L. et al. 1997. Biodiversity and ecosystem functioning in soil. *Ambio* 26: 563–570.

6. Snelgrove, P. R. et al. 1997. The importance of marine sediment biodiversity in ecosystem processes. *Ambio* 26: 578–583.

7. Wilson, E. O. 1997. The little things that run the world: the importance and conservation of invertebrates. *Conservation Biology* 1: 344–346.

8. Soil Quality Institute, U.S. Department of Agriculture, Soil Biology Quick Facts, online at http://soils.usda.gov/sqi/soil_quality/soil_biology/index.html.

9. Whitman, W. B., D. C. Coleman, and W. J. Wiebe. 1998. Prokaryotes: the unseen majority. *Proceedings of the National Academy of Sciences* 95: 6578–6583.

10. Gross, M. 1996. *Life on the Edge: Amazing Creatures Thriving in Extreme Environments*. Cambridge, MA: Perseus Books.

11. Treonis, A. M., D. H. Wall, and R. A. Virginia. 2000. The use of anhydrobiosis by soil nematodes in the Antarctic Dry Valleys. *Functional Ecology* 14: 460–467.

12. Smith, M. L., J. N. Bruhn, and J. B. Anderson. 1992. The fungus *Armillaria bulbosa* is among the largest and oldest living organisms. *Nature* 356: 428–431.

13. Ferguson, B. A. et al. 2003. Coarse-scale population structure of pathogenic *Armillaria* species in a mixed-conifer forest in the Blue Mountains of northeast Oregon. *Canadian Journal of Forestry Research* 33: 612–623.

14. Coleman, D. C. and D. A. Crossley Jr. 1996. *Fundamentals of Soil Ecology*. San Diego: Academic Press.

15. Wilson, The little things that run the world.

16. Wall, D. H., A. Fitter, and E. A. Paul. In press. Developing new perspectives from advances in soil biodiversity research. In *Biological Diversity and Function in Soils*, ed. R. D. Bardgett, M. B. Usher, and D. W. Hopkins. Cambridge, U.K.: Cambridge University Press.

17. Daily, G. C. 1995. Restoring value to the world's degraded lands. *Science* 269: 350–354.

18. Pimentel, D. et al. 1995. Environmental and economic costs of soil erosion and conservation benefits. *Science* 267: 1117–1123. But see also Kaiser, J. 2004. Wounding earth's fragile skin. *Science* 304: 1616–1618.

19. Wall, D. H., G. Adams, and A. N. Parsons. 2001. Soil biodiversity. In *Global Biodiversity in a Changing Environment: Scenarios for the 21st Century*, ed. F. S. Chapin III, O. E. Sala, and E. Huber-Sannwald, 47–82. Berlin: Springer-Verlag.

20. For example, the member nations of the Convention on Biological Diversity (CBD) decided at the 6th Conference of the Parties meeting

in Nairobi in April 2002 (COP decision VI/5, paragraph 13) "to establish an International Initiative for the Conservation and Sustainable Use of Soil Biodiversity as a cross-cutting initiative within the programme of work on agricultural biodiversity, and invites the Food and Agriculture Organization of the United Nations, and other relevant organizations, to facilitate and coordinate this initiative." See the CBD Web site at http://www.biodiv.org/default.aspx and the FAO Web site at http://www.fao.org/ag/agl/agll/soilbiod/fao.stm.

21. Baskin, Y. 1997. *The Work of Nature: How the Diversity of Life Sustains Us*. Washington, D.C.: Island Press; and Baskin, Y. 2002. *A Plague of Rats and Rubbervines: The Growing Threat of Species Invasions*. Washington, D.C.: Island Press/Shearwater Books.

22. Wall, D. H., ed. 2004. *Sustaining Biodiversity and Ecosystem Services in Soils and Sediments*. Washington, D.C.: Island Press.

23. Ward, P. D. and D. Brownlee. 2002. *The Life and Death of Planet Earth*. New York: Times Books.

24. Fortey, R. 1998. *Life: A Natural History of the First Four Billion Years of Life on Earth*. New York: Alfred A. Knopf. Quote p. 49.

25. Schwartzman, D. W. and T. Volk. 1989. Biotic enhancement of weathering and the habitability of Earth. *Nature* 340: 457–460.

26. Coleman and Crossley, *Fundamentals of Soil Ecology*.

27. Pimentel, D. et al. 1993. Soil erosion and agricultural productivity. In *World Soil Erosion and Conservation*, ed. D. Pimentel, 277–292. Cambridge, UK: Cambridge University Press.

28. Coleman and Crossley, *Fundamentals of Soil Ecology*.

29. Richter and Markewitz, How deep is soil?

II. Where Nematodes Are Lions

1. I visited Antarctica in the 2003–2004 season as a participant in the National Science Foundation's Antarctic Artists and Writers Program.

2. For an overview of the Taylor Valley ecosystems, see the six articles in a Special Section on the McMurdo Dry Valleys, *BioScience* 49 (12), December 1999.

3. Baldwin, J. G., S. A. Nadler, and D. H. Wall. 1999. Nematodes: Pervading the Earth and linking all life. In *Nature and Human Society: The Quest for a Sustainable World*, ed. P. H. Raven and T. Williams, 176–191. Washington, D.C.: National Academy Press.

4. Barrett, J. E. et al. 2004. Variation in biogeochemistry and soil biodiversity across spatial scales in a polar desert ecosystem. *Ecology* 85: 3105–3118.

5. Scott, R. F. 1905. *The Voyage of the Discovery*, Vol. II. London: Macmillan.

6. Dougherty, E. C. and L. G. Harris. 1963. Antarctic Micrometazoa: Fresh-water species in the McMurdo Sound area. *Science* 140: 497–498.

7. Horowitz, N. H. et al. 1969. Sterile soil from Antarctica: Organic analysis. *Science* 164: 1054–1056.

8. Horowitz, N. H. et al. 1972. Microbiology of the Dry Valleys of Antarctica. *Science* 176: 242–245. Quote p. 245.

9. Ezell, E. C. and L. N. Ezell. 1984. *On Mars: Exploration of the Red Planet 1958–1978*. The NASA History Series. Washington, D.C.: Scientific and Technical Information Branch, National Aeronautics and Space Administration. Online at http://www.solarviews.com/history/SP-4212/ch7-4.html; and Vishniac, W. V. and S. E. Mainzer. 1973. Antarctica as a Martian model. *Life Sciences and Space Research* 11: 25–31.

10. Friedmann, E. I. 1982. Endolithic microorganisms in the Antarctic cold desert. *Science* 215: 1045–1053; and Friedmann, E. I. and R. Ocampo. 1976. Endolithic blue-green algae in the Dry Valleys: Primary producers in the Antarctic desert ecosystem. *Science* 193: 1247–1249.

11. Freckman, D. W. and R. A. Virginia. 1997. Low-diversity Antarctic soil nematode communities: distribution and response to disturbance. *Ecology* 78: 363–369; and Freckman, D. W. and R. A. Virginia. 1989. Plant-feeding nematodes in deep-rooting desert ecosystems. *Ecology* 70: 1665–1678.

12. Ettema, C. H. and D. A. Wardle. 2002. Spatial soil ecology. *Trends in Ecology & Evolution* 17: 177–183.

13. Coleman, D. C. and D. A. Crossley Jr. 1996. *Fundamentals of Soil Ecology*. San Diego: Academic Press.

14. Virginia, R. A. and D. H. Wall. 1999. How soils structure communities in the Antarctic Dry Valleys. *BioScience* 49: 973–983.

15. Freckman, D. W. and R. A. Virginia. 1998. Soil biodiversity and community structure in the McMurdo Dry Valleys, Antarctica. American Geophysical Union *Antarctic Research Series* 72: 323–335.

16. Freckman, D. W. and R. A. Virginia. 1997. Low-diversity Antarctic soil nematode communities: distribution and response to disturbance. *Ecology* 78: 363–369.

17. Wall, D. H. and Virginia, R. A. 1999. The world beneath our feet: Soil biodiversity and ecosystem functioning. In *Nature and Human Society: The Quest for a Sustainable World*, ed. P. H. Raven and T. Williams, 225–241. Washington, D.C.: National Academy Press.

18. Baldwin, Nadler, and Wall, Nematodes: Pervading the Earth and linking all life.

19. Ibid.

20. Ibid.

21. Freckman and Virginia, Low-diversity Antarctic soil nematode communities.

22. Virginia and Wall, How soils structure communities in the Antarctic Dry Valleys.

23. Stevens, M. I. and I. D. Hogg. 2002. Expanded distributional records of Collembola and Acari in southern Victoria Land, Antarctica. *Pedobiologia* 46: 485–495; and Stevens, M. I. and I. D. Hogg. 2003. Long-term isolation and recent range expansion from glacial refugia revealed for the endemic springtail *Gomphiocephalus hodgsoni* from Victoria Land, Antarctica. *Molecular Ecology* 12: 2357–2369.

24. Treonis, A. M., D. H. Wall, and R. A. Virginia. 2000. The use of anhydrobiosis by soil nematodes in the Antarctic dry valleys. *Functional Ecology* 14: 460–467.

25. Wharton, D. A. 2002. *Life at the Limits: Organisms in Extreme Environments*. Cambridge, U.K.: Cambridge University Press.

26. Treonis, Wall, and Virginia, The use of anhydrobiosis by soil nematodes.

27. Courtright, E. M. et al. 2000. Nuclear and mitochondrial DNA sequence diversity in the Antarctic nematode *Scottnema lindsayae*. *Journal of Nematology* 32: 143–153.

28. Lal, R. 1999. World soils and the greenhouse effect. *The International Geosphere-Biosphere Programme, Global Change Newsletter* 37: 4–5.

29. Virginia and Wall, How soils structure communities in the Antarctic Dry Valleys.

30. Barrett et al., Variation in biogeochemistry and soil biodiversity.

31. Burkins, M. B. et al. 2000. Origin and distribution of soil organic matter in Taylor Valley, Antarctica. *Ecology* 81: 2377–2391.

32. Burkins, M. B., R. A. Virginia, and D. H. Wall. 2001. Organic carbon cycling in Taylor Valley, Antarctica: Quantifying soil reservoirs and soil respiration. *Global Change Biology* 7: 113–125.

33. Doran, P. T. et al. 2002. Antarctic climate cooling and terrestrial ecosystem response. *Nature* 415: 517–520.

34. Thompson, D. W. J. and S. Solomon. 2002. Interpretation of recent southern hemisphere climate change. *Science* 296: 895–899; Kerr, R. A. 2002. A single climate mover for Antarctica. *Science* 296: 825–826; Karoly, D. J. 2003. Ozone and climate change. *Science* 302: 236–237; and Gillett, N. P. and D. W. J. Thompson. 2003. Simulation of recent southern hemisphere climate change. *Science* 302: 273–275.

35. Barrett, J. E. et al. In preparation.

III. Of Ferns, Bears, and Slime Molds

1. For current air quality conditions and visibility in Great Smoky Mountains National Park, see http://www2.nature.nps.gov/air/webcams/parks/grsmcam/grsmcam.htm.

2. National Parks Conservation Association. 2004. *State of the Parks: Great Smoky Mountains National Park, A Resource Assessment.* See online at http://www.npca.org/stateoftheparks.

3. Renfro, J. 2002. Air pollution concerns at Great Smoky Mountains National Park. All Taxa Biodiversity Inventory *ATBI Quarterly* 3 (Spring): 6. Online at http://www.discoverlifeinamerica.org/.

4. Ibid.

5. See http://www.discoverlifeinamerica.org/.

6. Yoon, C. K. 1993. Counting creatures great and small. *Science* 260: 620–622.

7. Kaiser, J. 1997. Unique, all-taxa survey in Costa Rica "self-destructs." *Science* 276: 893; and see response: Gamez, R. et al. 1997. Costa Rican all-taxa survey. *Science* 277: 18–19.

8. Kaiser, J. 1997. Smoky Mountains all-taxa survey proposed. *Science* 278: 1871; and Sharkey, M. J. 2001. The all taxa biological inventory of the Great Smoky Mountains National Park. *Florida Entomologist* 84: 556–564.

9. Wardle, D. A. 2002. *Communities and Ecosystems: Linking the Aboveground and Belowground Components*. Monographs in Population Biology 34. Princeton: Princeton University Press.

10. Sharkey, The all taxa biological inventory of the Great Smoky Mountains National Park.

11. See http://www.discoverlifeinamerica.org/.

12. Wall, D. H., G. Adams, and A. N. Parsons. 2001. Soil biodiversity. In *Global Biodiversity in a Changing Environment: Scenarios for the 21st Century*, ed. F. S. Chapin III, O. E. Sala, and E. Huber-Sannwald, 47–82. Berlin: Springer-Verlag.

13. Arnolds, E. 1991. Decline of ectomycorrhizal fungi in Europe. *Agriculture, Ecosystems, and Environment* 35: 209–244.

14. Thomas, R. H. 2003. Wingless insects and plucked chickens. *Science* 299: 1854–1855.

15. Scheller, U. 2002. Pauropoda—the little ones among the Myriapods. All Taxa Biodiversity Inventory *ATBI Quarterly* 3 (Autumn): 3. Online at http://www.discoverlifeinamerica.org/.

16. Wall, D. H. and R. A. Virginia. 1999. The world beneath our feet: Soil biodiversity and ecosystem functioning. In *Nature and Human*

Society: The Quest for a Sustainable World, ed. P. H. Raven and T. Williams, 225–241. Washington, D.C.: National Academy Press.

17. Vogel, G. 2003. Royal Society: taxonomists endangered. *Science* 301: 153; and Wheeler, Q. D., P. H. Raven, and E. O. Wilson. 2004. Taxonomy: impediment or expedient? *Science* 303: 285.

18. Wilson, E. O. 2000. A global biodiversity map. *Science* 289: 2279.

19. The ALL Species Foundation's goal is to census all the world's species in 25 years. See Lawler, A. 2001. Up for the count? *Science* 294: 769–770.

20. See http://www.all-species.org/pbi.html.

21. Freckman, D. W. et al. 1997. Linking biodiversity and ecosystem functioning of soils and sediments. *Ambio* 26: 556–562.

22. Ettema, C. H. and D. A. Wardle. 2002. Spatial soil ecology. *Trends in Ecology & Evolution* 17: 177–183.

23. Soil Quality Institute, U.S. Department of Agriculture, Soil Biology Quick Facts, online at http://soils/usda.gov/sqi/soil_quality/soil_biology/index.html.

24. Wardle, *Communities and Ecosystems*.

25. Porazinska, D. L. et al. 2003. Relationships at the aboveground-belowground interface: plants, soil biota, and soil processes. *Ecological Monographs* 73: 377–395.

26. De Deyn, G. B. et al. 2004. Plant species identity and diversity effects on different trophic levels of nematodes in the soil food web. *Oikos* 106: 576–586.

27. Wardle, D. A. et al. 2004. Ecological linkages between aboveground and belowground biota. *Science* 304: 1629–1633; and Wardle, *Communities and Ecosystems*.

28. Kowalchuk, G. A. et al. 2002. Effects of above-ground plant species composition and diversity on the diversity of soil-borne microorganisms. *Antonie van Leeuwenhoek* 81: 509–520.

29. Kowalchuk, G. A. et al. 2000. Changes in the community structure of ammonia-oxidizing bacteria during secondary succession of calcareous grasslands. *Environmental Microbiology* 2: 99–110.

30. Van der Putten, W. H. 2003. Plant defense belowground and spatio-

temporal processes in natural vegetation. *Ecology* 84: 2269–2280; and Wardle, *Communities and Ecosystems.*

31. De Deyn, G. B. et al. 2003. Soil invertebrate fauna enhances grassland succession and diversity. *Nature* 422: 711–713.

32. Laakso, J. and H. Setala. 1999. Sensitivity of primary production to changes in the architecture of belowground food webs. *Oikos* 87: 57–64.

33. Loreau, M. et al. 2001. Biodiversity and ecosystem functioning: current knowledge and future challenges. *Science* 294: 804–808.

34. Symstad, A. J. et al. 2003. Long-term and large-scale perspectives on the relationship between biodiversity and ecosystem functioning. *BioScience* 53: 89–98; and Wardle, *Communities and Ecosystems.*

35. Heemsbergen, D. A. et al. 2004. Biodiversity effects on soil processes explained by interspecific functional dissimilarity. *Science* 306: 1019–1020; and Wardle et al., Ecological linkages between aboveground and belowground biota.

36. See Global Litter Invertebrate Decomposition Experiment (GLIDE) online at http://www.nrel.colostate.edu/projects/glide/index.html.

37. Freckman et al., Linking biodiversity and ecosystem functioning of soils and sediments.

38. Brussaard, L. et al. 1997. Biodiversity and ecosystem functioning in soil. *Ambio* 26: 563–570.

39. Schimel, J. 1995. Ecosystem consequences of microbial diversity and community structure. In *Arctic and Alpine Biodiversity: Patterns, Causes and Ecosystem Consequences*, ed. F. S. Chapin III and C. Korner, 239–254. Berlin: Springer-Verlag. Quote p. 250.

40. Wall, Adams, and Parsons, Soil biodiversity.

IV. The Power of Ecosystem Engineers

1. Darwin, C. 1881. *The Formation of Vegetable Mould Through the Action of Worms with Observations on Their Habits.* London: Murray. Quote p. 4.

2. Ibid. Quote p. 6.

3. Brown, G. G., C. A. Edwards, and L. Brussaard. 2004. How earth-worms affect plant growth: burrowing into the mechanisms. In *Earthworm Ecology*, 2nd ed., ed. C. A. Edwards, 13–49. Boca Raton: CRC Press.

4. White, G. 1789. *The Natural History of Selborne*. London: Benjamin White. (Quotes from Penguin Classics 1987 edition, p. 196.)

5. World Wildlife Fund Traveling Exhibit, "Biodiversity 911: Saving Life on Earth."

6. Lavelle, P. et al. 1997. Soil function in a changing world: the role of invertebrate ecosystem engineers. *European Journal of Soil Biology* 33: 159–193.

7. Jones, C. G., J. H. Lawton, and M. Shachak. 1994. Organisms as ecosystem engineers. *Oikos* 69: 373–386.

8. Darwin, *The Formation of Vegetable Mould Through the Action of Worms*.

9. Lavelle et al., Soil function in a changing world.

10. Visit Minnesota Worm Watch online at http://www.nrri.umn.edu/worms.

11. Nixon, W. 1995. As the worm turns. *American Forests* 101: 34–36.

12. For findings attributed to Hale, see: Hale, C. M. 2004. Ecological consequences of exotic invaders: Interactions involving European earthworms and native plant communities in hardwood forests. Ph.D. Dissertation, University of Minnesota, Department of Forest Resources, St. Paul, Minnesota.

13. Gundale, M. J. 2002. Influence of exotic earthworms on the soil or-ganic horizon and the rare fern *Botrychium mormo*. *Conservation Biology* 16: 1555–1561.

14. Chauvel, A. et al. 1999. Pasture damage by an Amazonian earth-worm. *Nature* 398: 32–33; and Lavelle et al., Soil function in a changing world.

15. Groffman, P. M. and P. J. Bohlen. 1999. Soil and sediment biodiver-sity: cross-system comparisons and large-scale effects. *BioScience* 49: 139–148.

16. Soil Quality Institute, U.S. Department of Agriculture, The Soil Biology Primer, online at http://soils.usda.gov/sqi/soil_quality/soil_biology/soil_biology_primer.html.

17. Dominguez, J., P. J. Bohlen, and R. W. Parmelee. 2004. Earthworms increase nitrogen leaching to greater soil depths in row crop agroecosystems. *Ecosystems* 7: 672–685.

18. Bohlen, P. J. et al. 2004. Ecosystem consequences of exotic earthworm invasion of north temperate forests. *Ecosystems* 7: 1–12.

19. Hendrix, P. F. and P. J. Bohlen. 2002. Exotic earthworm invasions in North America: Ecological and policy implications. *BioScience* 52: 801–811.

20. Bohlen et al., Ecosystem consequences of exotic earthworm invasion.

21. Burtelow, A. E., P. J. Bohlen, and P. M. Groffman. 1998. Influence of exotic earthworm invasion on soil organic matter, microbial biomass, and denitrification potential in forest soils of the northeastern United States. *Applied Soil Ecology* 9: 197–202.

22. Buech, R. R., C. P. Scissions, and D. J. Rugg. In preparation. Effect of exotic earthworms on a forest small-mammal community.

23. Maerz, J. C. et al. 2005. Introduced invertebrates are important prey for a generalist predator. *Diversity and Distributions* 11: 83–90.

24. Simberloff, D. and B. Von Holle. 1999. Positive interactions of non-indigenous species: invasional meltdown? *Biological Invasions* 1: 21–32. For discussion of synergy among invasive pigs, earthworms, etc., in Hawaii, see Baskin, Y. 2002. *A Plague of Rats and Rubbervines: The Growing Threat of Species Invasions*. Washington, D.C.: Island Press/Shearwater Books.

25. Kourtev, P. S., W. Z. Huang, and J. G. Ehrenfeld. 1999. Differences in earthworm densities and nitrogen dynamics in soils under exotic and native plant species. *Biological Invasions* 1: 237–245.

26. Heneghan, L. 2003. And when they got together: the impacts of Eurasian earthworm and invasive shrubs on Chicago woodland ecosystems. *Chicago Wilderness Journal* 1: 27–31; and Heneghan, L. et al. 2004. European buckthorn (*Rhamnus cathartica*) and its effects on some ecosystem properties in urban woodland. *Ecological Restoration* 22: 275–280.

27. Bohlen, P. J. et al. 2004. Non-native invasive earthworms as agents of change in northern temperate forests. *Frontiers in Ecology and Environment* 2: 427–435.

28. Stein, A. et al. 1992. Spatial variability of earthworm populations in a permanent polder grassland. *Biology and Fertility of Soils* 14: 260–266; and Hoogerkamp, M., H. Rogaar, and H. J. P. Eijsackers. 1983. Effect of earthworms on grassland on recently reclaimed polder soils in the Netherlands. In *Earthworm Ecology: From Darwin to Vermiculture*, ed. J. E. Satchell, 85–105. London: Chapman and Hall.

29. Stockdill, S. M. J. 1982. Effects of introduced earthworms on the productivity of New Zealand pastures. *Pedobiologia* 24: 29–35.

30. Brown, G. G. et al. 1999. Effects of earthworms on plant production in the tropics. In *Earthworm Management in Tropical Agroecosystems*, ed. P. Lavelle, L. Brussaard, and P. F. Hendrix, 87–147. Wallingford, U.K.: CAB International.

31. Senapati, B. K. et al. 1999. In-soil technologies for tropical ecosystems. In *Earthworm Management in Tropical Agroecosystems*, ed. P. Lavelle, L. Brussaard, and P. F. Hendrix, 199–237. Wallingford, U.K.: CAB International; and Senapati, B. K. et al. 2002. Restoring soil fertility and enhancing productivity in Indian tea plantations with earthworms and organic fertilizers. In *International Technical Workshop on Biological Management of Soil Ecosystems for Sustainable Agriculture: Program, Abstracts and Related Documents*, 24–27 June 2002, 172–190. Londrina, Brazil: Embrapa Soybean. Online at http://www.fao.org/ag/agl/agll/soilbiod/cases/casea1.doc.

32. Folgarait, P. J. 1998. Ant biodiversity and its relationship to ecosystem functioning: a review. *Biodiversity and Conservation* 7: 1221–1244; and Lavelle et al., Soil function in a changing world.

33. Hölldobler, B. and E. O. Wilson. 1990. *The Ants*. Cambridge, MA: The Belknap Press of Harvard University Press. Quote p. 1.

34. Sanchez, P. A. et al. In press. *Alternatives to Slash-and-Burn: A Global Synthesis*. Madison, WI: American Society of Agronomy Special Publication; and Giller, K. E. et al. In press. Soil biodiversity in rapidly changing tropical landscapes: scaling down and scaling up. In *Biological Diversity and Function in Soils*, ed. R. D. Bardgett,

M. B. Usher, and D. W. Hopkins. Cambridge, U.K.: Cambridge University Press.

35. Mando, A. et al. 2002. Managing termites and organic resources to improve soil productivity in the Sahel. In *International Technical Workshop on Biological Management of Soil Ecosystems for Sustainable Agriculture: Program, Abstracts and Related Documents*, 24–27 June 2002, 191–203. Londrina, Brazil: Embrapa Soybean. Online at http://www.fao.org/ag/agl/agll/soilbiod/cases/casea2_n.doc.

v. Plowing the Seabed

1. Snelgrove, P. R. 1999. Getting to the bottom of marine biodiversity: Sedimentary habitats. *BioScience* 49: 129–138.

2. Snelgrove, P. R. et al. 1997. The importance of marine sediment biodiversity in ecosystem processes. *Ambio* 26: 578–583.

3. Malakoff, D. 2003. Scientists counting on census to reveal marine biodiversity. *Science* 302: 773. Online at http://www.coml.org.

4. Thrush, S. F. and P. K. Dayton. 2002. Disturbance to marine benthic habitats by trawling and dredging: Implications for marine biodiversity. *Annual Review of Ecology and Systematics* 33: 449–473.

5. Snelgrove et al., The importance of marine sediment biodiversity.

6. Thrush and Dayton, Disturbance to marine benthic habitats.

7. Snelgrove et al., The importance of marine sediment biodiversity.

8. Andersen, F. O. and E. Kristensen. 1991. Effects of burrowing macrofauna on organic matter decomposition in coastal marine sediments. *Symposium of The Zoological Society of London* 63: 69–88.

9. Hutchings, P. 1998. Biodiversity and functioning of polychaetes in benthic sediments. *Biodiversity and Conservation* 7: 1133–1145.

10. Smith, C. R. et al. 2000. Global change and biodiversity linkages across the sediment-water interface. *BioScience* 50: 1108–1120.

11. National Research Council. 2002. *Effects of Trawling and Dredging on Seafloor Habitat*. Washington, D.C.: National Academy Press.

12. Watling, L. and E. A. Norse. 1998. Disturbance of the sea bed by mobile fishing gear: A comparison to forest clearcutting. *Conservation Biology* 12: 1180–1197.

13. Smith et al., Global change and biodiversity linkages.

14. Roberts, C. M. 2002. Deep impact: The rising toll of fishing in the deep sea. *Trends in Ecology & Evolution* 17: 242–245.

15. Snelgrove, Getting to the bottom of marine biodiversity.

16. Levin, L. A. et al. 2001. Environmental influences on regional deep-sea species diversity. *Annual Review of Ecology and Systematics* 32: 51–93. But see also Poore, G. C. B. et al. 1997. Coastal and deep-sea benthic diversities compared. *Marine Ecology Progress Series* 159: 97–103.

17. Snelgrove et al., The importance of marine sediment biodiversity.

18. Smith et al., Global change and biodiversity linkages.

19. Watling and Norse, Disturbance of the sea bed by mobile fishing gear.

20. National Research Council, *Effects of Trawling and Dredging on Seafloor Habitat*; and Collie, J. S. et al. 2000. A quantitative analysis of fishing impacts on shelf-sea benthos. *Journal of Animal Ecology* 69: 785–798.

21. Jennings, S. et al. 2002. Effects of chronic trawling disturbance on the production of infaunal communities. *Marine Ecology Progress Series* 243: 251–262.

22. National Research Council, *Effects of Trawling and Dredging on Seafloor Habitat*.

23. Online at http://www.cost-impact.org.

24. Kaiser, M. J. et al. In review. Recovery rates of benthos and habitats subjected to fishing disturbance. *Marine Ecology Progress Series*.

25. Collie et al., A quantitative analysis of fishing impacts.

26. Austen, M. C. et al. In preparation. Effects of fishing on nutrient cycling.

27. Townsend, M. et al. In preparation. Relative importance of bioturbation for nutrient cycling in sand and muddy habitats.

28. Reichelt, A. C. 1991. Environmental effects of meiofaunal burrowing. *Symposium of The Zoological Society of London* 63: 33–52.

29. Schratzberger, M., T. A. Dinmore, and S. Jennings. 2002. Impacts of trawling on the biodiversity, biomass, and structure of meiofauna assemblages. *Marine Biology* 140: 83–93; and Schratzberger, M. and S. Jennings. 2002. Impacts of chronic trawling disturbance on meiofaunal communities. *Marine Biology* 141: 991–1000.

30. Widdicombe, S. et al. 2004. The importance of bioturbators for biodiversity maintenance: The indirect effects of fishing disturbance. *Marine Ecology Progress Series* 275: 1–10.

31. Widdicombe, S. et al. 2000. Bioturbation as a mechanism for setting and maintaining levels of diversity in subtidal macrobenthic communities. *Hydrobiologia* 440: 369–377; Widdicombe, S. and M. C. Austen. 1999. Mesocosm investigation into the effects of bioturbation on the diversity and structure of a subtidal macrobenthic community. *Marine Ecology Progress Series* 189: 181–193.

32. Austen, M. C. and S. Widdicombe. 1998. Experimental evidence of effects of the heart urchin *Brissopsis lyrifera* on associated subtidal meiobenthic nematode communities. *Journal of Experimental Marine Biology and Ecology* 222: 219–238.

33. Widdicombe et al., The importance of bioturbators for biodiversity maintenance.

34. Thayer, C. W. 1979. Biological bulldozers and the evolution of marine benthic communities. *Science* 203: 458–461; Thayer, C. W. 1983. Sediment-mediated biological disturbance and the evolution of marine benthos. In *Biotic Interactions in Recent and Fossil Benthic Communities*, ed. M. J. S. Tevesz, 479–625. New York: Plenum.

35. Costanza, R. et al. 1997. The value of the world's ecosystem services and natural capital. *Nature* 387: 253–260.

36. National Research Council, *Effects of Trawling and Dredging on Seafloor Habitat*.

37. Roberts, S. and M. Hirshfield. 2004. Deep-sea corals: out of sight, but no longer out of mind. *Frontiers in Ecology & the Environment* 2: 123–130.

38. United Nations Environment Programme. 2004. *GEO Year Book 2003*. Online at http://www.unep.org/geo/yearbook.

VI. Microbes, Muck, and Dead Zones

1. Rabalais, N. N., R. E. Turner, and D. Scavia. 2002. Beyond science into policy: Gulf of Mexico hypoxia and the Mississippi River. *BioScience* 52: 129–142.

2. National Research Council. 2000. *Clean Coastal Waters: Understanding and Reducing the Effects of Nutrient Pollution*. Washington, D.C.: National Academy Press. Online at http://www.nap.edu/catalog/9812.html.

3. United Nations Environment Programme. 2004. *GEO Year Book 2003*. Online at http://www.unep.org/geo/yearbook.

4. Howarth, R. W. et al. 1996. Regional nitrogen budgets and riverine N & P fluxes for the drainages to the North Atlantic Ocean: natural and human influences. *Biogeochemistry* 35: 75–139.

5. Ibid.

6. Galloway, J. N. et al. 2003. The nitrogen cascade. *BioScience* 53: 341–356.

7. Mitsch, W. J. et al. 2001. Reducing nitrogen loading to the Gulf of Mexico from the Mississippi River Basin: Strategies to counter a persistent ecological problem. *BioScience* 51: 373–388; and National Research Council. 1992. *Restoration of Aquatic Ecosystems: Science, Technology, and Public Policy*. Washington, D.C.: National Academy Press. Online at http://books.nap.edu/catalog/1807.html.

8. Rabalais, Turner, and Scavia, Beyond science into policy.

9. National Research Council, *Restoration of Aquatic Ecosystems*.

10. Odum, E. P. 1971. *Fundamentals of Ecology*, 3rd ed. Philadelphia: W. B. Saunders Co.; and Odum, H. T. 1971. *Environment, Power, and Society*. New York: Wiley-Interscience.

11. Mitsch, W. J. 1996. Ecological engineering: a new paradigm for engineers and ecologists. In *Engineering within Ecological Constraints*, ed. P. C. Schulze, 111–128. Washington, D.C.: National Academy Press. Online at http://www.nap.edu/catalog/4919.html.

12. Rabalais, Turner, and Scavia, Beyond science into policy.

13. Levin, L. A. 2002. Deep-ocean life where oxygen is scarce. *American Scientist* 90: 436–444; and Rabalais, Turner, and Scavia, Beyond science into policy.

14. Rabalais, Turner, and Scavia, Beyond science into policy.

15. Turner, R. E. and N. N. Rabalais. 2003. Linking landscape and water quality in the Mississippi River basin for 200 years. *BioScience* 53: 563–572.

16. Ferber, D. 2001. Keeping the Stygian waters at bay. *Science* 291: 968–973.

17. Diaz, R. J. and A. Solow. 1999. *Ecological and Economic Consequences of Hypoxia: Topic 2 Report for the Integrated Assessment on Hypoxia in the Gulf of Mexico.* Silver Spring, MD: NOAA Coastal Ocean Office. Decision Analysis Series no. 19. Online at http://www.cop.noaa.gov/pubs/das/das16.pdf.

18. Ibid.; and National Research Council, *Clean Coastal Waters.*

19. Levin, L. A. et al. 2001. The function of marine critical transition zones and the importance of sediment biodiversity. *Ecosystems* 4: 430–451.

20. Peterson, C. H. and J. Lubchenco. 1997. Marine ecosystem services. In *Nature's Services: Societal Dependence on Natural Ecosystems,* ed. G. C. Daily, 177–194. Washington, D.C.: Island Press.

21. Levin et al., The function of marine critical transition zones; and Widdicombe, S. et al. 2004. The importance of bioturbators for biodiversity maintenance: The indirect effects of fishing disturbance. *Marine Ecology Progress Series* 275: 1–10.

22. Diaz and Solow, *Ecological and Economic Consequences of Hypoxia.*

23. Committee on Environment and Natural Resources. 2000. *Integrated Assessment of Hypoxia in the Northern Gulf of Mexico.* Washington, D.C.: National Science and Technology Council. Online at http://www.nos.noaa.gov/Products/pubs_hypox.html.

24. Rabalais, Turner, and Scavia, Beyond science into policy.

25. Howarth et al., Regional nitrogen budgets.

26. National Research Council, *Clean Coastal Waters.*

27. Mississippi River/Gulf of Mexico Watershed Nutrient Task Force. 2001. *Action Plan for Reducing, Mitigating, and Controlling Hypoxia in the Northern Gulf of Mexico.* Washington, D.C.: Office of

Wetlands, Oceans, and Watersheds, U.S. Environmental Protection Agency.

28. Turner and Rabalais, Linking landscape and water quality.

29. Mitsch, W. J. et al. 1999. *Reducing Nutrient Loads, Especially Nitrate-Nitrogen, to Surface Water, Ground Water, and the Gulf of Mexico: Topic 5 Report for the Integrated Assessment on Hypoxia in the Gulf of Mexico.* Silver Spring, MD: NOAA Coastal Ocean Office. Decision Analysis Series no. 19. Online at http://www.cop.noaa.gov/pubs/das/das19.pdf; and Mitsch et al., Reducing nitrogen loading to the Gulf of Mexico.

30. Precise definitions of "wetland" continue to shift in both scientific and legal/regulatory usage. See National Research Council. 1995. *Wetlands: Characteristics and Boundaries.* Washington, D.C.: National Academy Press. Online at http://books.nap.edu/catalog/4766.html.

31. Meyer, O. 1994. Functional groups of microorganisms. In *Biodiversity and Ecosystem Function*, ed. E.-D. Schulze and H. A. Mooney, 67–96. Berlin: Springer-Verlag.

32. Mitsch, Ecological engineering: A new paradigm for engineers and ecologists.

33. Mitsch, W. J. and R. F. Wilson. 1996. Improving the success of wetland creation and restoration with know-how, time, and self-design. *Ecological Applications* 6: 77–83; and Mitsch, W. J. et al. 1998. Creating and restoring wetlands. *BioScience* 48: 1019–1030.

34. DeWeerdt, S. 2004. Reflections on the pond. *Conservation in Practice* 5: 21–27.

35. Mitsch et al., Creating and restoring wetlands.

36. Turner and Rabalais, Linking landscape and water quality.

37. Howarth et al., Regional nitrogen budgets.

38. Mitsch, W. J., and J. G. Gosselink. 2000. The value of wetlands: Importance of scale and landscape setting. *Ecological Economics* 35: 25–33.

39. Hey, D. L. and N. S. Philippi. 1995. Flood reduction through wetland restoration: The Upper Mississippi River Basin as a case history. *Restoration Ecology* 3: 4–17.

40. Mitsch et al., *Reducing Nutrient Loads, Especially Nitrate-Nitrogen*.

41. Mitsch, W. J. et al. In press. Nitrate-nitrogen retention by wetlands in the Mississippi River Basin. *Ecological Engineering*.

42. Ibid.

43. Ewel, K. C. 1997. Water quality improvement by wetlands. In *Nature's Services: Societal Dependence on Natural Ecosystems*, ed. G. C. Daily, 329–344. Washington, D.C.: Island Press; and Kadlec, R. and R. Knight. 1996. *Treatment Wetlands*. Boca Raton, FL: CRC Press.

44. Mitsch et al., *Reducing Nutrient Loads, Especially Nitrate-Nitrogen*.

45. Mitsch et al., Nitrate-nitrogen retention by wetlands.

46. Mitsch et al., Reducing nitrogen loading to the Gulf of Mexico.

47. Turner and Rabalais, Linking landscape and water quality.

48. Mitsch et al., Reducing nitrogen loading to the Gulf of Mexico.

49. Diaz, R. J., J. Nestlerode, and M. L. Diaz. 2004. A global perspective on the effects of eutrophication and hypoxia on aquatic biota. In *Proceedings of the 7th International Symposium on Fish Physiology, Toxicology, and Water Quality*, Tallinn, Estonia, 12–15 May 2003, ed. G. L. Rupp and M. D. White, 1–33. Athens, GA: U.S. Environmental Protection Agency. EPA 600/R-04/049.

VII. Fungi and the Fate of Forests

1. MacKinnon, A. and J. A. Trofymow. 1998. Structure, processes, and diversity in successional forests of coastal British Columbia. In *Proceedings of a Workshop on Structure, Process, and Diversity in Successional Forests of Coastal British Columbia*, ed. J. A. Trofymow and A. MacKinnon, 1–3. *Northwest Science* 72 (Special Issue No. 2); and Schoonmaker, P. K., B. von Hagen, and E. C. Wolf, ed. 1997. *The Rain Forests of Home*. Washington, D.C.: Island Press.

2. Norse, E. A. 1990. *Ancient Forests of the Pacific Northwest*. Washington, D.C.: Island Press.

3. See, for instance, Dietrich, W. 1992. *The Final Forest: The Battle for the Last Great Trees of the Pacific Northwest*. New York: Simon & Schuster; Bunnell, F. and L. Kremsater. 2003. Resolving forest

management issues in British Columbia. In *Towards Forest Sustainability*, ed. D. B. Lindenmayer and J. F. Franklin, 85–110. Washington, D.C.: Island Press; and Coady, L. (B.C. Coastal Group, Weyerhaeuser) 2002. New approaches to environmental challenges in British Columbia's coastal forests. Presented at the Interforest Congress, Munich, Germany, July 2002. Online at http://www.coastforestconservationinitiative.com/news_updates/events_and_presentations.html.

4. Kohm, K. A. and J. F. Franklin. 1997. *Creating a Forestry for the 21st Century: The Science of Ecosystem Management*. Washington, D.C.: Island Press.

5. Franklin, J. F. et al. 1997. Alternative silvicultural approaches to timber harvesting: Variable retention harvest systems. In *Creating a Forestry for the 21st Century: The Science of Ecosystem Management*, ed. K. A. Kohm and J. F. Franklin, 111–139. Washington, D.C.: Island Press.

6. Perry, D. A. 1994. *Forest Ecosystems*. Baltimore: The Johns Hopkins University Press.

7. Pilz, D. et al. 2003. *Ecology and Management of Commercially Harvested Chanterelle Mushrooms*. General Technical Report PNW-GTR-576. Portland, OR: U.S. Department of Agriculture, Forest Service, Pacific Northwest Research Station.

8. Perry, *Forest Ecosystems*.

9. Palmer, J. D. and S. L. Baldauf. 1993. Animals and fungi are each other's closest relatives: Congruent evidence from multiple proteins. *Proceedings of the National Academy of Sciences* 90: 11,558–11,562.

10. Perry, *Forest Ecosystems*; and Schowalter, T. et al. 1997. Integrating the ecological roles of phytophagous insects, plant pathogens, and mycorrhizae in managed forests. In *Creating a Forestry for the 21st Century: The Science of Ecosystem Management*, ed. K. A. Kohm and J. F. Franklin, 171–189. Washington, D.C.: Island Press.

11. Franklin, J. F. et al. 1985. Ecosystem responses to the eruption of Mount St. Helens. *National Geographic Research* (Spring): 198–216.

12. Perry, D. A. 2003. The role of science in the changing forestry scene in the USA. In *Towards Forest Sustainability*, ed. D. B. Lindenmayer

and J. F. Franklin, 127–141. Washington, D.C.: Island Press; and Perry, D. A. and M. P. Amaranthus. 1997. Disturbance, recovery, and stability. In *Creating a Forestry for the 21st Century: The Science of Ecosystem Management*, ed. K. A. Kohm and J. F. Franklin, 31–56. Washington, D.C.: Island Press.

13. Perry and Amaranthus, Disturbance, recovery, and stability.

14. Franklin, J. F. 1989. Toward a new forestry. *American Forests* (Nov./Dec.): 1–8.

15. See Weyerhaeuser's Coast Forest Strategy online at http://cacr.forestry. ubc.ca/forest_strategy/default.htm.

16. Kohm, K. A. and J. F. Franklin. 1997. Introduction. In *Creating a Forestry for the 21st Century: The Science of Ecosystem Management*, ed. K. A. Kohm and J. F. Franklin, 1–5. Washington, D.C.: Island Press. Quote p. 5.

17. See Weyerhaeuser's Coast Forest Strategy, Introduction, online at http://cacr.forestry.ubc.ca/forest_strategy/intro/.

18. Bunnell, F. et al. 2003. *Learning to Sustain Biological Diversity on Weyerhaeuser's Coastal Tenure*. Nanaimo, British Columbia: Weyerhaeuser.

19. Outerbridge, R. A. and J. A. Trofymow. 2004. Diversity of ectomycorrhizae on experimentally planted Douglas fir seedlings in variable retention forestry sites on southern Vancouver Island. *Canadian Journal of Botany* 82: 1671–1681.

20. Franklin, J. F. et al. 1999. The history of DEMO: An experiment in regeneration harvest of northwestern forest ecosystems. *Northwest Science* 73 (Special Issue): 3–11.

21. Luoma, D. L. et al. 2004. Response of ectomycorrhizal fungus sporocarp production to varying levels and patterns of green-tree retention. *Forest Ecology and Management* 202: 337–354.

22. Trofymow, J. A. and G. L. Porter. 1998. Introduction to the coastal forest chronosequence project. In *Proceedings of a Workshop on Structure, Process, and Diversity in Successional Forests of Coastal British Columbia*, ed. J. A. Trofymow and A. MacKinnon, 4–8. *Northwest Science* 72 (Special Issue No. 2).

23. MacKinnon and Trofymow, Structure, processes, and diversity in successional forests.

24. Trofymow, J. A. et al. 2003. Attributes and indicators of old growth and successional Douglas fir forests on Vancouver Island. *Environmental Reviews* 11 (Issue Supplement 1): S187–S204.

25. Goodman, D. M. and J. A. Trofymow. 1998. Comparison of communities of ectomycorrhizal fungi in old-growth and mature stands of Douglas fir at two sites on southern Vancouver Island. *Canadian Journal of Forestry Research* 28: 574–581.

26. Ibid.

27. Goodman, D. M., J. A. Trofymow, and A. J. Thomson. 2000. Developing an online database of descriptions of ectomycorrhizae. *British Columbia Journal of Ecosystems and Management* 1: 1–8. Online database at http://www.pfc.forestry.ca/biodiversity/bcern/index_e.html.

28. Marcot, B. G. 1997. Biodiversity of old forests of the West: A lesson from our elders. In *Creating a Forestry for the 21st Century: The Science of Ecosystem Management*, ed. K. A. Kohm and J. F. Franklin, 87–105. Washington, D.C.: Island Press.

29. Perry, *Forest Ecosystems*.

30. Perry and Amaranthus, Disturbance, recovery, and stability; and Perry, *Forest Ecosystems*.

31. Ibid.

32. Ineson, P. et al. 2004. Cascading effects of deforestation on ecosystem services across soil and freshwater and marine sediments. In *Sustaining Biodiversity and Ecosystem Services in Soils and Sediments*, ed. D. H. Wall, 225–248. Washington, D.C.: Island Press.

33. Bunnell and Kremsater, Resolving forest management issues in British Columbia.

34. Trofymow et al., Attributes and indicators of old growth.

VIII. Grazers, Grass, and Microbes

1. Huff, D. E. and J. D. Varley. 1999. Natural regulation in Yellowstone National Park's northern range. *Ecological Applications* 9: 17–29; and Coughenour, M. B. and F. J. Singer. 1991. The concept of over-

grazing and its application to Yellowstone's Northern Range. In *The Greater Yellowstone Ecosystem: Redefining America's Wilderness Heritage*, ed. R. B. Keiter and M. S. Boyce, 209–230. New Haven: Yale University Press.

2. Frank, D. A., S. J. McNaughton, and B. F. Tracy. 1998. The ecology of the Earth's grazing ecosystems. *BioScience* 48: 513–521.

3. McNaughton, S. J. 1985. Ecology of a grazing ecosystem: The Serengeti. *Ecological Monographs* 55: 259–294; and Frank, McNaughton, and Tracy, The ecology of the Earth's grazing ecosystems.

4. Frank, D. A. and P. M. Groffman. 1998. Ungulate vs. landscape control of soil C and N processes in grasslands of Yellowstone National Park. *Ecology* 79: 2229–2241.

5. Frank, D. A. and S. J. McNaughton. 1993. Evidence for the promotion of aboveground grassland production by native large herbivores in Yellowstone National Park. *Oecologia* 96: 157–161.

6. Bardgett, R. D. and D. A. Wardle. 2003. Herbivore-mediated linkages between aboveground and belowground communities. *Ecology* 84: 2258–2268; and Wardle, D. A. 2002. *Communities and Ecosystems: Linking the Aboveground and Belowground Components.* Monographs in Population Biology 34. Princeton: Princeton University Press.

7. Wardle, D. A. and R. D. Bardgett. 2004. Human-induced changes in large herbivorous mammal density: The consequences for decomposers. *Frontiers in Ecology and the Environment* 2: 145–153.

8. Frank, D. A. and S. J. McNaughton. 1992. The ecology of plants, large mammalian herbivores, and drought in Yellowstone National Park. *Ecology* 73: 2043–2058; and Frank, McNaughton, and Tracy, The ecology of the Earth's grazing ecosystems.

9. McNaughton, S. J. 1988. Mineral nutrition and spatial concentrations of African ungulates. *Nature* 334: 343–345; McNaughton, S. J. 1990. Mineral nutrition and seasonal movements of African migratory ungulates. *Nature* 345: 613–615; and Frank, McNaughton, and Tracy, The ecology of the Earth's grazing ecosystems.

10. Frank, McNaughton, and Tracy, The ecology of the Earth's grazing ecosystems.

11. National Research Council. 2002. *Ecological Dynamics on Yellowstone's Northern Range*. Washington, D.C.: National Academy Press. Online at http://www.nap.edu/catalog/10328.html.

12. Dyksterhuis, E. J. 1949. Condition and management of rangeland based on quantitative ecology. *Journal of Range Management* 2: 104–105.

13. Briske, D. D., S. D. Fuhlendorf, and F. E. Smeins. 2003. Vegetation dynamics on rangelands: A critique of the current paradigms. *Journal of Applied Ecology* 40: 601–614.

14. Coughenour and Singer, The concept of overgrazing.

15. National Research Council, *Ecological Dynamics on Yellowstone's Northern Range*. Quote p. 104.

16. Ibid. Quote p. 103.

17. Smith, D. W., R. O. Peterson, and D. B. Houston. 2003. Yellowstone after wolves. *BioScience* 53: 330–340; and Beschta, R. L. 2003. Cottonwoods, elk, and wolves in the Lamar Valley of Yellowstone National Park. *Ecological Applications* 13: 1295–1309.

18. Hamilton, E. W. and D. A. Frank. 2001. Can plants stimulate soil microbes and their own nutrient supply? Evidence from a grazing tolerant grass. *Ecology* 82: 2397–2402.

19. Vitousek, P. M. and R. W. Howarth. 1991. Nitrogen limitation on land and in the sea: How can it occur? *Biogeochemistry* 13: 87–115.

20. Holland, J. N., W. Cheng, and D. A. Crossley Jr. 1996. Herbivore-induced changes in plant carbon allocation: Assessment of below-ground C fluxes using carbon-14. *Oecologia* 107: 87–94; Dyer, M. I. and U. G. Bokhari. 1976. Plant-animal interactions: Studies of the effects of grasshopper grazing on blue grama grass. *Ecology* 57: 762–772; and Bardgett, R. D., D. A. Wardle, and G. W. Yeates. 1998. Linking above-ground and below-ground interactions: How plant responses to foliar herbivory influence soil organisms. *Soil Biology and Biochemistry* 30: 1867–1878.

21. Denton, C. S. et al. 1999. Low amounts of root herbivory positively influence the rhizosphere microbial community of a temperate grassland soil. *Soil Biology and Biochemistry* 31: 155–165; and Yeates, G. W. et al. 1998. Impact of clover cyst nematode (*Heterodera trifolii*) infection on soil microbial activity in the rhizosphere of white clover

(*Trifolium repens*)—a pulse labelling experiment. *Nematologica* 44: 81–90.

22. Holland, J. N. 1995. Effects of aboveground herbivory on soil microbial biomass in conventional and no-tillage agroecosystems. *Applied Soil Ecology* 2: 275–279; and Mawdsley, J. L. and Bardgett, R. D. 1997. Continuous defoliation of perennial ryegrass (*Lolium perenne*) and white clover (*Trifolium repens*) and associated changes in the microbial population of an upland soil. *Biology and Fertility of Soils* 24: 52–58.

23. Merrill, E. H., N. L. Stanton, and J. C. Hak. 1994. Responses of bluebunch wheatgrass, Idaho fescue, and nematodes to ungulate grazing in Yellowstone National Park. *Oikos* 69: 231–240.

24. Mikola, J. et al. 2001. Effects of defoliation intensity on soil food-web properties in an experimental grassland community. *Oikos* 92: 333–343.

25. Ingham, R. E. et al. 1985. Interactions of bacteria, fungi, and their nematode grazers: Effects on nutrient cycling and plant growth. *Ecological Monographs* 55: 119–140; and Verhoef, H. A. and L. Brussaard. 1990. Decomposition and nitrogen mineralization in natural and agroecosystems: the contribution of soil animals. *Biogeochemistry* 11: 175–211.

26. Coleman, D. C. and D. A. Crossley Jr. 1996. *Fundamentals of Soil Ecology*. San Diego: Academic Press.

27. Frank, D. A., M. M. Kuns, and D. R. Guido. 2002. Consumer control of grassland plant production. *Ecology* 83: 602–606.

28. Frank, D. A. et al. 2003. Soil community composition and the regulation of grazed temperate grassland. *Oecologia* 137: 603–609.

29. Ibid.

30. Gehring, C. A. and T. G. Whitham. 1994. Interactions between aboveground herbivores and the mycorrhizal mutualists of plants. *Trends in Ecology and Evolution* 9: 251–255. But see also Wallace, L. L. 1981. Growth, morphology and gas exchange of mycorrhizal and nonmycorrhizal *Panicum coloratum* L., a C4 grass species, under different clipping and fertilization regimes. *Oecologia* 49: 272–278.

31. Frank et al., Soil community composition.

32. Van der Putten, W. H. and B. A. M. Peters. 1997. How soil-borne pathogens may affect plant competition. *Ecology* 78: 1785–1795; Van der Putten, W. H., C. Van Dijk, and B. A. M. Peters. 1993. Plant-specific soil-borne diseases contribute to succession in foredune vegetation. *Nature* 362: 53–56; Packer, C. and K. Clay. 2000. Soil pathogens and spatial patterns of seedling mortality in a temperate tree. *Nature* 404: 278–281; and Klironomos, J. N. 2002. Feedback with soil biota contributes to plant rarity and invasiveness in communities. *Nature* 417: 67–70.

33. Van der Heijden, M. G. A. et al. 1998. Mycorrhizal fungal diversity determines plant biodiversity, ecosystem variability, and productivity. *Nature* 396: 69–72; and Bever, J. D. 2002. Host-specificity of AM fungal population growth rates can generate feedback on plant growth. *Plant and Soil* 244: 281–290.

34. Bever, J. D. 2003. Soil community feedback and the coexistence of competitors: Conceptual frameworks and empirical tests. *New Phytologist* 157: 465–473.

35. Frank, McNaughton, and Tracy, The ecology of the Earth's grazing ecosystems; and Frank, D. A. 1998. Ungulate regulation of ecosystem processes in Yellowstone National Park: Direct and feedback effects. *Wildlife Society Bulletin* 26: 410–418.

36. Bardgett and Wardle, Herbivore-mediated linkages; and Wardle, *Communities and Ecosystems.*

37. Van der Wal, R. et al. 2004. Vertebrate herbivores and ecosystem control: cascading effects of faeces on tundra ecosystems. *Ecography* 27: 242–252.

38. Collins, S. L. et al. 1998. Modulation of diversity by grazing and mowing in native tallgrass prairie. *Science* 280: 745–747; and Johnson, L. C. and J. R. Matchett. 2001. Fire and grazing regulate belowground processes in tallgrass prairie. *Ecology* 82: 3377–3389.

39. Milchunas, D. G. and W. K. Lauenroth. 1993. Quantitative effects of grazing on vegetation and soils over a global range of environments. *Ecological Monographs* 63: 327–366.

40. Pastor, J. et al. 1988. Moose, microbes, and the boreal forest. *BioScience* 38: 770–777.

41. Bardgett and Wardle, Herbivore-mediated linkages; and Wardle and

Bardgett, Human-induced changes in large herbivorous mammal density.

42. Wardle, D. A. et al. 2001. Introduced browsing mammals in New Zealand natural forests: Aboveground and belowground consequences. *Ecological Monographs* 71: 587–614.

43. Bardgett and Wardle, Herbivore-mediated linkages.

44. Olff, H., M. E. Ritchie, and H. H. T. Prins. 2002. Global environmental controls of diversity in large herbivores. *Nature* 415: 901–904.

IX. Restoring Power to the Soil

1. Swift, M. J. et al. 1996. Biodiversity and agroecosystem function. In *Functional Roles of Biodiversity: A Global Perspective*, ed. H. A. Mooney et al., 261–298. Chichester, U.K.: John Wiley & Sons Ltd.

2. Matson, P. A. et al. 1997. Agricultural intensification and ecosystem properties. *Science* 277: 504–509.

3. Mikola, J., R. D. Bardgett, and K. Hedlund. 2002. Biodiversity, ecosystem functioning, and soil decomposer food webs. In *Biodiversity and Ecosystem Functioning: Synthesis and Perspectives*, ed. M. Loreau, S. Naeem, and P. Inchausti, 169–180. New York: Oxford University Press.

4. Lal, R. et al. 2004. Managing soil carbon. *Science* 304: 393; and Lal, R. 2004. Soil carbon sequestration impacts on global climate change and food security. *Science* 304: 1623–1627.

5. Mikola, Bardgett, and Hedlund, Biodiversity, ecosystem functioning, and soil decomposer food webs; and Swift et al., Biodiversity and agroecosystem function.

6. Buckley, D. H. and T. M. Schmidt. 2001. The structure of microbial communities in soil and the lasting impact of cultivation. *Microbial Ecology* 42: 11–21.

7. Mikola, Bardgett, and Hedlund, Biodiversity, ecosystem functioning, and soil decomposer food webs; and Siepel, H. 1996. Biodiversity of soil microarthropods: The filtering of species. *Biodiversity and Conservation* 5: 251–260.

8. Matson, Agricultural intensification and ecosystem properties.

9. Daily, G. C. 1995. Restoring value to the world's degraded lands. *Science* 269: 350–354.

10. Food and Agriculture Organization of the United Nations. 2000. Sustainable land use and management needed to prevent soil degradation. Press release, 4 May 2000. Online media archive at http://www.fao.org/WAICENT.

11. Food and Agriculture Organization of the United Nations. 1998. Soil protection needs more attention in Europe. Press release, 26 May 1998. Online media archive at http://www.fao.org/WAICENT.

12. Soil and Water Conservation Society. 1999. Soil—A critical environmental resource. Soil Fact Sheet Packet. Available online at http://swcs.org/en/publications/books/soil_fact_sheet_packet.cfm.

13. McNeill, J. R. and V. Winiwarter. 2004. Breaking the sod: Humankind, history, and soil. *Science* 304: 1627–1629. Quote p. 1628.

14. Huston, M. 1993. Biological diversity, soils, and economics. *Science* 262: 1676–1680.

15. Robertson, H. J. and R. G. Jefferson. 2002. Nature and farming in Britain. In *The Farm as Natural Habitat: Reconnecting Food Systems with Ecosystems*, ed. D. L. Jackson and L. L. Jackson, 123–135. Washington, D.C.: Island Press.

16. Ibid.

17. Thomas, J. A. et al. 2004. Comparative losses of British butterflies, birds, and plants and the global extinction crisis. *Science* 303: 1879–1881.

18. Kleijn, D. et al. 2001. Agri-environment schemes do not effectively protect biodiversity in Dutch agricultural landscapes. *Nature* 413: 723–725; Kleijn, D. and W. J. Sutherland. 2003. How effective are European agri-environment schemes in conserving and promoting biodiversity? *Journal of Applied Ecology* 40: 947–969; and Kleijn, D. et al. 2004. Ecological effectiveness of agri-environment schemes in different agricultural landscapes in the Netherlands. *Conservation Biology* 18: 775–786.

19. Gibson, C. W. D. and V. K. Brown. 1991. The nature and rate of development of calcareous grasslands in Southern England. *Biological Conservation* 58: 297–316.

20. Including CLUE (Changing Land Use, Enhancement of Biodiversity and Ecosystem Development) and TLinks (Trophic Linkages Between Above- and Below-Ground Organisms). Online at http://www.nioo.knaw.nl/cto/clue/clue.htm and http://www.bf.jcu.cz/tlinks/, respectively.

21. Van der Putten, W. H. et al. 2000. Plant species diversity as a driver of early succession in abandoned fields: a multi-site approach. *Oecologia* 124: 91–99.

22. Hedlund, K. et al. 2003. Plant species diversity, plant biomass, and responses of the soil community on abandoned land across Europe: Idiosyncrasy or above-belowground time lags. *Oikos* 103: 45–58.

23. Arnolds, E. 1991. Decline of ectomycorrhizal fungi in Europe. *Agriculture, Ecosystems, and Environment* 35: 209–244; and Wall, D. H., G. Adams, and A. N. Parsons. 2001. Soil biodiversity. In *Global Biodiversity in a Changing Environment: Scenarios for the 21st Century*, ed. F. S. Chapin III, O. E. Sala, and E. Huber-Sannwald, 47–82. Berlin: Springer-Verlag.

24. Stevens, C. J. et al. 2004. Impact of nitrogen deposition on the species richness of grasslands. *Science* 203: 1876–1879.

25. Kaiser, J. 2001. The other global pollutant: Nitrogen proves tough to curb. *Science* 294: 1268–1269.

26. De Deyn, G. B. et al. 2003. Soil invertebrate fauna enhances grassland succession and diversity. *Nature* 422: 711–713.

27. De Deyn, G. B., C. E. Raaijmakers, and W. H. Van der Putten. 2004. Plant community development is affected by nutrients and soil biota. *Journal of Ecology* 92: 786–796.

28. Bever, J. D. 2003. Soil community feedback and the coexistence of competitors: conceptual frameworks and empirical tests. *New Phytologist* 157: 465–473.

29. Van der Heijden, M. G. A. et al. 1998. Mycorrhizal fungal diversity determines plant biodiversity, ecosystem variability, and productivity. *Nature* 396: 69–72; and Bever, Soil community feedback.

30. Van der Putten, W., C. Van Dijk, and B. A. M. Peters. 1993. Plant-specific soil-borne diseases contribute to succession in foredune vegetation. *Nature* 362: 53–56.

31. Van der Putten, W. H. et al. 2001. Linking above- and belowground multitrophic interactions of plants, herbivores, pathogens, and their antagonists. *Trends in Ecology & Evolution* 16: 547–554.

32. Gange, A. C. and V. K. Brown. 2002. Actions and interactions of soil invertebrates and arbuscular mycorrhizal fungi in affecting the structure of plant communities. In *Mycorrhizal Ecology*, Ecological Studies, Vol. 157, ed. M. G. A. van der Heijden and I. Sanders, 321–344. Berlin: Springer-Verlag.

33. Van der Putten, W. H. 2003. Plant defense belowground and spatiotemporal processes in natural vegetation. *Ecology* 84: 2269–2280; and Bezemer, T. M. et al. 2004. Above- and belowground terpenoid aldehyde induction in cotton, *Gossypium herbaceum*, following root and leaf injury. *Journal of Chemical Ecology* 30: 53–67.

34. Bezemer, T. M. et al. In review. What regulates an invasive plant species in its native area? Interplay between plant community diversity, negative plant-soil feedbacks, and aboveground herbivores. *Ecological Monographs*.

35. De Deyn, G. B. et al. 2004. Plant species identity and diversity effects on different trophic levels of nematodes in the soil food web. *Oikos* 106: 576–586.

36. Swift et al., Biodiversity and agroecosystem function.

37. Myers, N. 1999. Pushed to the edge: The fates of rainforest wildlife and marginal farmers are intertwined. *Natural History* (March): 20–22; Fox, J. et al. 2000. Shifting cultivation: A new old paradigm for managing tropical forests. *BioScience* 50: 521–528; and Myers, N. 1993. Tropical forests: The main deforestation fronts. *Environmental Conservation* 20: 9–16.

38. See the United Nations Food and Agriculture Organization Soil Biodiversity Portal at http://www.fao.org/ag/agl/agll/soilbiod/fao.stm and the Convention on Biological Diversity agricultural biodiversity site at http://www.biodiv.org/programmes/areas/agro/.

39. Food and Agriculture Organization of the United Nations. 2003. *Biological Management of Soil Ecosystems for Sustainable Agriculture*, World Soil Resources Report 101, Report of the International Technical Workshop organized by Embrapa Soybean and FAO, Londrina, Brazil, 24–27 June 2002. Quote p. 1. Online at http://www.fao.org/documents/show_cdr.asp?url_file-/docrep/006/y4810e/y4810e00.htm.

40. Environment News Service. 2002. Soil's tiniest organisms could solve huge problems. November 29. Online at http://www.ens-newswire.com/ens/nov2002/2002-11-29-02.asp.

41. See Conservation and Sustainable Management of Below-Ground Biodiversity online at http://www.ciat.cgiar.org/tsbf_institute/csm_bgbd.htm.

42. See Alternatives to Slash-and-Burn Programme online at http://www.asb.cgiar.org/home.htm; and Bignell, D. E. et al. In press. Below-ground biodiversity assessment: The ASB rapid, functional group approach. In *Alternatives to Slash-and-Burn: A Global Synthesis*, ed. P. A. Sanchez et al. Madison, WI: American Society of Agronomy Special Publication.

43. Bignell et al., Below-ground biodiversity assessment.

44. Giller, K. E. et al. In press. Soil biodiversity in rapidly changing tropical landscapes: Scaling down and scaling up. In *Biological Diversity and Function in Soils*, ed. R. D. Bardgett, M. B. Usher, and D. W. Hopkins. Cambridge, U.K.: Cambridge University Press.

Epilogue

1. Ward, P. D. and D. Brownlee. 2002. *The Life and Death of Planet Earth*. New York: Times Books.

2. Brussaard, L. et al. 1997. Biodiversity and ecosystem functioning in soil. *Ambio* 26: 563–570; Snelgrove, P. R. et al. 1997. The importance of marine sediment biodiversity in ecosystem processes. *Ambio* 26: 578–583; and Klironomos, J. N. 2002. Another form of bias in conservation research. *Science* 298: 749.

3. See *Science*. 2004. Special Section: Soils—The Final Frontier. 304: 1613–1637.

4. Environmental News Service. 2002. Soil's tiniest organisms could solve huge problems. November 29. Online at http://www.ens-newswire.com/ens/nov2002/2002-11-29-02.asp.

5. Copley, J. 2000. Ecology goes underground. *Nature* 406: 452–454.

6. Kowalchuk, G. A., M. Bruinsma, and J. A. van Veen. 2003. Assessing response of soil microorganisms to GM plants. *Trends in Ecology and Evolution* 18: 403–410.

7. Amundson, R., Y. Guo, and P. Gong. 2003. Soil diversity and land use in the United States. *Ecosystems* 6: 470–482.

8. Wilson, E. O. 2002. *The Future of Life*. New York: Alfred A. Knopf. Quote pp. 145–146.

Acknowledgments

This book is an outgrowth of the SCOPE Soil and Sediment Biodiversity and Ecosystem Functioning (SSBEF) project, and I am fundamentally indebted to project leader Diana H. Wall and the dozens of scientists worldwide who participated in a series of SSBEF workshops between 1996 and 2002. Their efforts first introduced me to the vibrant and vital life of the soil and sediments and sparked my interest in telling this story. Support for the research and writing of this book was graciously provided by SCOPE and by the Winslow Foundation, which also helped to fund many of the interdisciplinary SSBEF workshops. I want to express deep appreciation to SCOPE executive director Veronique Plocq-Fichelet and editor in chief John W. B. Stewart for making this book possible.

Special thanks are also due to the National Science Foundation Antarctic Artists and Writers Program and its manager, Guy G. Guthridge, for making it possible for me to travel to the McMurdo Dry Valleys Long Term Ecological Research site with soil scientists during the 2003–2004 austral summer. Thanks also to the Soil Ecology Society and the British Ecological Society for welcoming me to their 2003 meetings, and to Dave Robins for helping to make possible my visit to Plymouth Marine Laboratory.

I'm especially indebted to the many scientists who provided professional, logistic, and often personal help, along with field tours, interviews, discussions, and reviews in the course of my research and travels. In the United Kingdom, Melanie C. Austen, Richard D. Bardgett, David E. Bignell, Valerie K. Brown, Michelle T. Fountain, Clare Lawson, Simon R. Mortimer, Simon Potts, and Duncan Westbury. In the Netherlands, Lijbert Brussaard, Gerlinde B. de Deyn, Ken E. Giller, Jeffrey A. Harvey, Paul Kardol, George A. Kowalchuk, Nicole M. van Dam, Wim H. Van der Putten, and Jasper van Ruijven. In New Zealand, David A. Wardle. In Brazil, George G. Brown. In France, Patrick Lavelle. In Canada, Jan A. Addison, Valerie M. Behan-Pelletier, Glen Dunsworth, Renata Outerbridge, and J. A. Trofymow. At McMurdo Station and in the McMurdo Dry Valleys of Antarctica, Byron J. Adams, John E. Barrett, Emma J. Broos, Laurie B. Connell, Scott D. Craig, David Hopkins, Diane M. McKnight and the "Stream Team," Johnson Nkem, Regina S. Redman, Russell J. Rodriguez, Ross A. Virginia, and Diana H. Wall, as well as Rae Spain, the helicopter crews, and the excellent staff of the Crary Science Laboratory at McMurdo, and Elaine Hood in Denver. And in the United States, Ernest C. Bernard, Patrick J. Bohlen, Nina F. Caraco, David C. Coleman, Katherine C. Ewel, Douglas A. Frank, Jerry F. Franklin, Lee E. Frelich, Cindy M. Hale, Paul Hendrix, Jeanie Hilten, Keith Langdon, Lisa A. Levin, Daniel L. Luoma, William J. Mitsch, Chuck Parker, David A. Perry, Steve Stephenson and members of the Slime Mold TWIG, and James M. Tiedje.

Dozens of other scientists and students shared their time and insights with me, although in many cases, space limitations kept me from using directly the materials they provided. I thank all of them and hope they recognize that their contributions helped to shape this book.

Despite all the invaluable advice and assistance I received, any errors or deficiencies, as well as the opinions and interpretations expressed in this book, are solely my responsibility.

Finally, I want to thank Jonathan Cobb, executive editor at Shearwater Books, for his enthusiastic support of this book from the beginning and his insightful editing of the final manuscript. And, as always, love and thanks to my husband Mike Gilpin for his daily support.

Index

Adams, Byron, 18–19, 27, 30–31
agriculture, intensive: agri-environment schemes to reduce impacts of, 169–71; changes in British countryside caused by, 167–70; erosion and degradation of soil by, 7–8, 166; impoverishes soil community, 164–65; and loss of plant diversity in grasslands, 168–69; overrides soil services, 164–66; restoration complicated by soil legacy of, 166, 171–73. *See also* restoration of former agricultural land
agriculture, tropical: Alternatives to Slash-and-Burn (ASB), 185; Conservation and Sustainable Management of Below-Ground Biodiversity (BGBD), 185–86; impacts on soil ecosystem engineers, 185; linking soil biodiversity to sustainability of, 78–79, 184–86; shifting cultivation and deforestation, 183–84; "soil biological management" in, 184–85

algae: as Antarctic carbon producers, 33–34, 36; in marine nutrient cycle, 84, 92; role in eutrophication, 101
All Taxa Biodiversity Inventory (ATBI): in Costa Rica, 42–43; in Great Smoky Mountains National Park, 40–41, 43; new initiatives, 49; value of cataloging species, 40, 44. *See also* Census of Marine Life; taxonomy
Alternatives to Slash-and-Burn (ASB), 185
Amaranthus, Mike, 140
anhydrobiosis, 29–30. *See also* cryptobiotic states
Antarctica. *See* dry valleys
Antarctic Treaty, 20
ants, as ecosystem engineers, 78, 185
Austen, Melanie, 80–82, 85–88, 91–99

bacteria: abundance of, 4; anaerobic processes of, 111; denitrifying, 111–15; factors fostering diversity of, 51–53; grazing influences

Day, John, 111, 117–19
"dead zones" (hypoxic waters)
101–2; and natural "oxygen minimum zones," 108. *See also* Gulf of
Mexico
decomposition. *See* nutrient cycling
de Deyn, Gerlinde, 179–80, 183
denitrification: in marine sediments,
95; and nitrous oxide release, 112,
114, 119; in wetland sediments
103, 111–15, 117–19
Discover Life in America, Inc., 41,
47. *See also* All Taxa Biodiversity
Inventory; Great Smoky Mountains
National Park
dredging. *See* trawl fishing
dry valleys, McMurdo (Antarctica),
14–15, 18, 21; carbon production
pulses and "legacy carbon" in,
33–34, 36; climate changes in,
35–37; Long Term Ecological Research (LTER) program in, 34;
map of, 16; as model for Mars, 22;
as oasis for life, 15, 21; polygon-
patterned ground in, 14, 19, 23;
soil biological activity in, 28–29,
36–37; soil nematodes in, 25–27,
31; sterility theory of arid soils in,
22, 25. *See also* nematodes
Dunsworth, Glen, 131–32

Earth, role of life in making it habitable, 10–12
earthworms: basic categories of, 70;
changes in forest floor, vegetation,
and animal life caused by invading,
64–69, 72, 74–75; as creators and
modifiers of soil habitat, 60–63;
Darwin's study of, 58–59, 62; deer
compounding effects of, 68, 76; impacts of intensive agricultural practices on, 164–65, 185; impacts of
plant diversity on, 175; increase in

soil compaction by, 68–69; introductions of exotic, 59–60, 69,
73–74; invading in Chippewa National Forest, Minnesota, 63–65,
70; nitrate leaching enhanced by,
73; nutrient cycling altered by,
65–66, 72–73; regions without native, 59; species-specific impacts of,
72–74; swings in reputation of, 59;
used to enhance plant growth,
77–78
ecological engineering, 107, 115
ecosystem engineers, 51; earthworms
as, 62–63; impact of tropical agricultural practices on, 185–86; in marine
sediments, 82; termites and ants as,
78–79. *See also* bioturbators
ecosystem services ("life support services"): consequences of soil species
diversity or loss for, 17, 24, 37,
55–57, 140–41; economic value of,
97; feedbacks of soil and sediment
organisms to climate change 191–92;
of fungi in forests, 124–25, 127;
impact of marine sediment organisms on nutrient cycling, 82–85,
91, 93–95; role of soil and sediment organisms in, 6–7, 40; of soil
organisms in maintaining soil fertility, 184–85; of wetland microbes in
recycling nitrogen, 102–3, 111–15
elk. *See* grazing
erosion, land degradation and, 7–8,
166
eutrophication, 82; and dead zones in
coastal waters, 101–2

Ferris, Bob, 122, 127, 135
Food and Agriculture Organization
(FAO), United Nations, 166, 184
forest floor (duff, mor humus), 65
forestry practices: adaptive management of, 130–32, 140; and

nutrient cycling, 152–56; stimulates grass and root production, 144–45, 152, 155–56

Great Smoky Mountains National Park: air pollution damage in, 39, 47; All Taxa Biodiversity Inventory in, 40–41, 44; biodiversity levels in, 43–44; human pressures on, 39–40; mushrooms and mycorrhizae in, 42, 47; slime molds in, 43–44; threats from invasive species in, 39, 41–42, 74. *See also* Discover Life in America, Inc.

Gulf of Mexico: action plan to reduce "dead zone" (hypoxia) in, 110; definition of hypoxia in, 108; extent of eutrophication and hypoxia in, 101, 107–11; extent of watershed draining to, 100–101; nitrate runoff from farmland to, 101, 109–10; potential damage to biodiversity and fisheries in, 108–9; stratification of water in, 108; wetland loss in Mississippi River basin draining to, 103, 105–6; wetland restoration needed to reduce nitrate runoff to, 110–11, 117–20. *See also* nitrogen

Hale, Cindy, 59–60, 63–73, 75–76
Hamilton, Bill, 152–55
Hendrix, Paul, 73
Hilten, Jeanie, 47
Hölldobler, Bert, 78
Hopkins, David, 52
humus, 33, 58; mor (forest floor, duff), 65; reduced by intensive agriculture, 165

invasive nonnative species: of earthworms, 59–60, 69, 73–74; threatening Great Smoky Mountains National Park, 39, 41–42; as threat to marine sediment life, 81. *See also* earthworms

Janzen, Dan, 42–43

Kowalchuk, George, 52, 54

Langdon, Keith, 38
Lavelle, Patrick, 62–63, 69
Lawson, Clare, 172
lichens: in Antarctic rocks, 22; in temperate rain forests, 123–24, 134, 136–37
Luoma, Daniel, 133

marram grass, 181
Mars: Antarctic dry valleys as model for, 22; prospect of life on, 1–3, 22; "soil" on, 2–3
Massulik, Stacey, 158
McNaughton, Sam, 144, 146–47
microbes: extremophiles, 4; influence on plant community, 180–81; and nutrient cycling in soils and sediments, 84–85; and role in making Earth habitable, 10–12, 188. *See also* bacteria; fungi
millipedes, 78, 164, 185
Mississippi River watershed, 100–101; wetland loss in, 103, 105–6. *See also* Gulf of Mexico
mites, 21, 27, 140, 165, 175; influence on plant succession, 179–80
Mitsch, William J., 104–7, 109–11, 115–20
mollusks (clams, scallops, snails), 46–47, 82, 85–88, 94
Mortimer, Simon, 167–73, 175–76
Murray, Tanya, 159
mushrooms. *See* chanterelles; fungi
mycorrhizae. *See* fungi

"Nematode National Park," 37, 193
nematodes, 16–17, 26, 81; anhydrobiosis in, 29–30; Antarctic habitat for, 19, 25; Antarctic species of, 25–27; in Chihuahuan desert, 23;

protozoa, 21; and nutrient cycling, 94, 155
protura, 48
Pseudomonas (bacteria), 113–14

Rabalais, Nancy, 107
rain forests: temperate coastal, 121–22, 134; tropical, 184. *See also* forestry practices
range science, changing ideas in, 148–49. *See also* grazing
restoration, wetland. *See* wetlands
restoration of former agricultural land: agri-environment schemes for, 169–71; requires reducing fertility (nitrogen), 168, 174, 177–78; soil legacies of cultivation slow, 166, 171–73; soil provides clues to success in, 171–72; strategies to speed soil recovery, plant succession during, 173–76, 178–83
rhizosphere, 24, 165
Richardson, Kirsten, 82, 86
Richter, Daniel, 2
rotifers, 21, 27; anhydrobiosis in, 29

salal shrubs, 123, 125
Scheller, Ulf, 48
SCOPE (Scientific Committee on Problems of the Environment) Soil and Sediment Biodiversity and Ecosystem Functioning project, 8–9
Scott, Robert Falcon, 21
Scottnema lindsayae (nematode), 26–27, 31, 36–37; drawing of, 28.
sediment organisms, freshwater: diversity of 3–4; nitrogen transformations by microbial, 111–15
sediment organisms, marine: diversity of, 3–4, 81, 89; impacts of bioturbator loss on diversity of, 96; impacts of eutrophication and hypoxia on, 108–09; impacts of trawl fishing on,

89–91, 93; and influence on nutrient cycling, 82–85, 91, 93–95; threats to, 7–8, 81–82, 90
Senapati, Bikram, 77
Serengeti plain: 98–99, 142–44, 146–47. *See also* grazing
Shadis, Dave, 64
shifting cultivation (slash and burn), 183–84
shrimp, burrowing, 82, 85–86; impacts of trawl fishing on, 91; role in marine nutrient cycling, 94
slime molds, 45–46, 50; global inventory of, 49–50; role in soil food webs, 43–44; Taxonomic Working Group (TWIG) for, 41, 43–44
snails. *See* mollusks
soil: carbon losses during intensive agriculture, 165; carbon stored in, 32–33; classification of, 12; constituents of, 12; definition of , 2–3, 12; diversity of life in, 3–5; erosion and degradation of, 7–8, 166; forest-derived, 184; formation of, 11–12; habitat diversity in, 23–24, 51–53; on Mars, 2–3; rare and endangered "series" of, 193; richer plant diversity on infertile, 168
soil ecology, growing interest and research in, 3, 6–8, 50–51, 188–89, 191–92
soil health and quality, international concern for, 8, 184–85
soil organisms: diversity of, 3–5, 43–44; dormancy among, 29–30, 51–53; ecological services of, 6–7, 40; and feedbacks to climate change, 191–92; functional groups of, 51; "functional redundancy" among, 56–57; grazed plants stimulate activity of, 152–56; impacts of forestry practices on, 128, 132–34, 139–40; impacts of intensive agriculture

131–32, 140; Coast Forest Strategy, 129, 139; sustainable forestry certification for, 131

White, Gilbert, 59

Widdicombe, Stephen, 96

Wilson, Edward O., 4, 6, 49, 78, 193

wireworms (click beetle larvae), 179–80

Wormherders, 16. *See also* dry valleys

Yellowstone National Park, 142–50. *See also* grazing